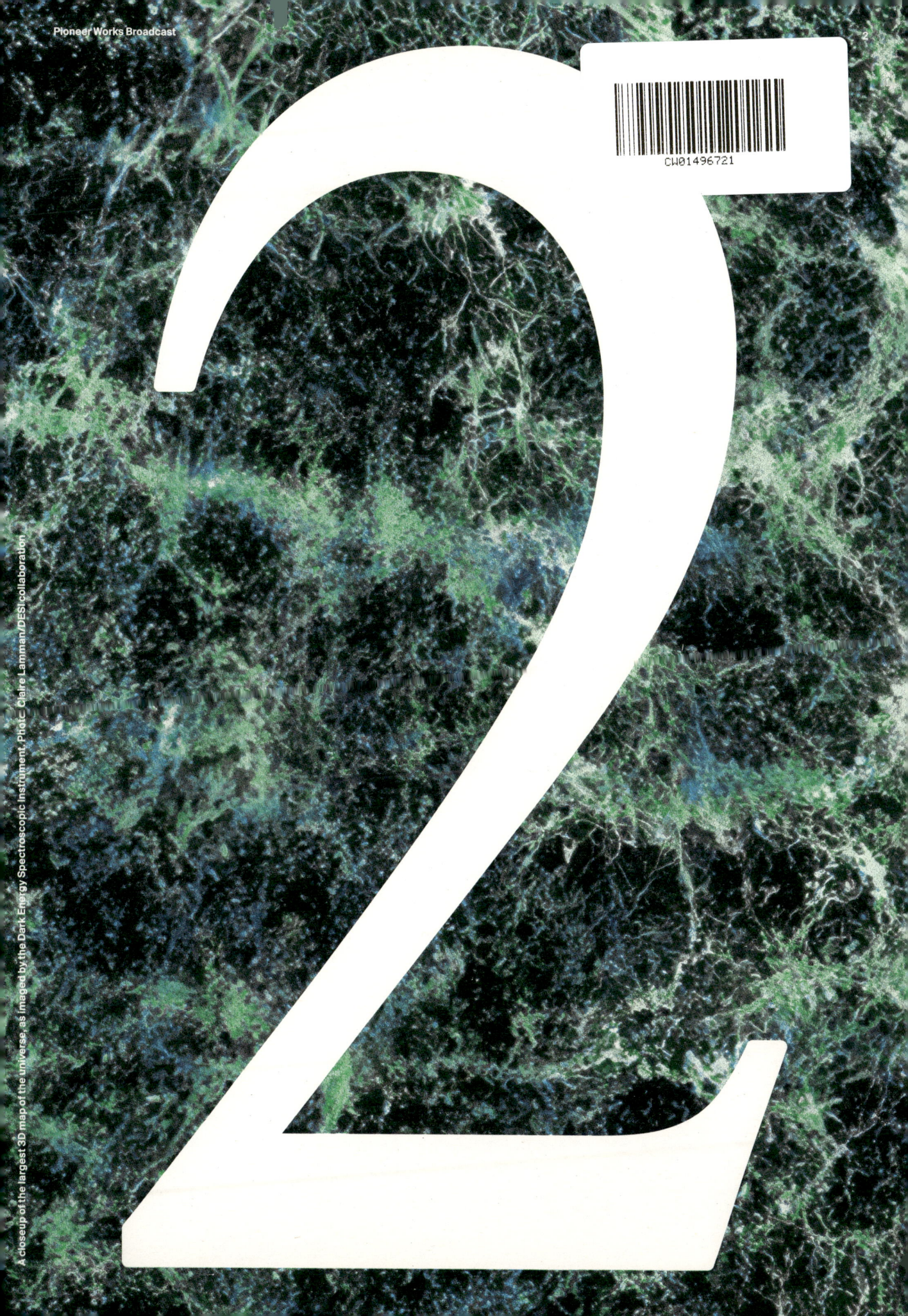

A closeup of the largest 3D map of the universe, as imaged by the Dark Energy Spectroscopic Instrument. Photo: Claire Lamman/*DESI* collaboration

¿Quién no ha intentado convertir una piedra en un recuerdo?
Alejandro García Contreras

September 6 – December 15, 2024
Pioneer Works

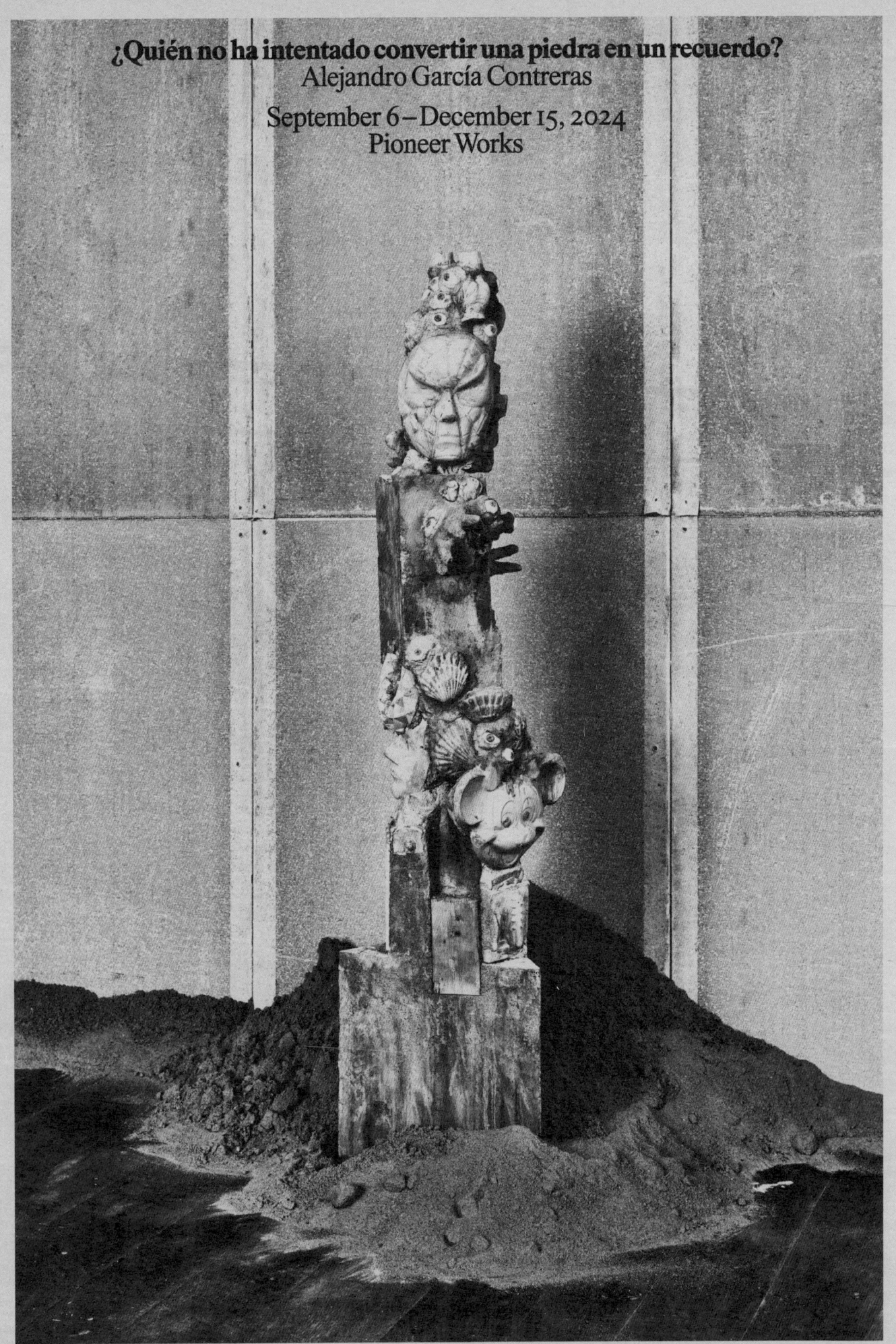

¿Quién no ha intentado convertir una piedra en un recuerdo?
Alejandro García Contreras

TANGENT SPACE

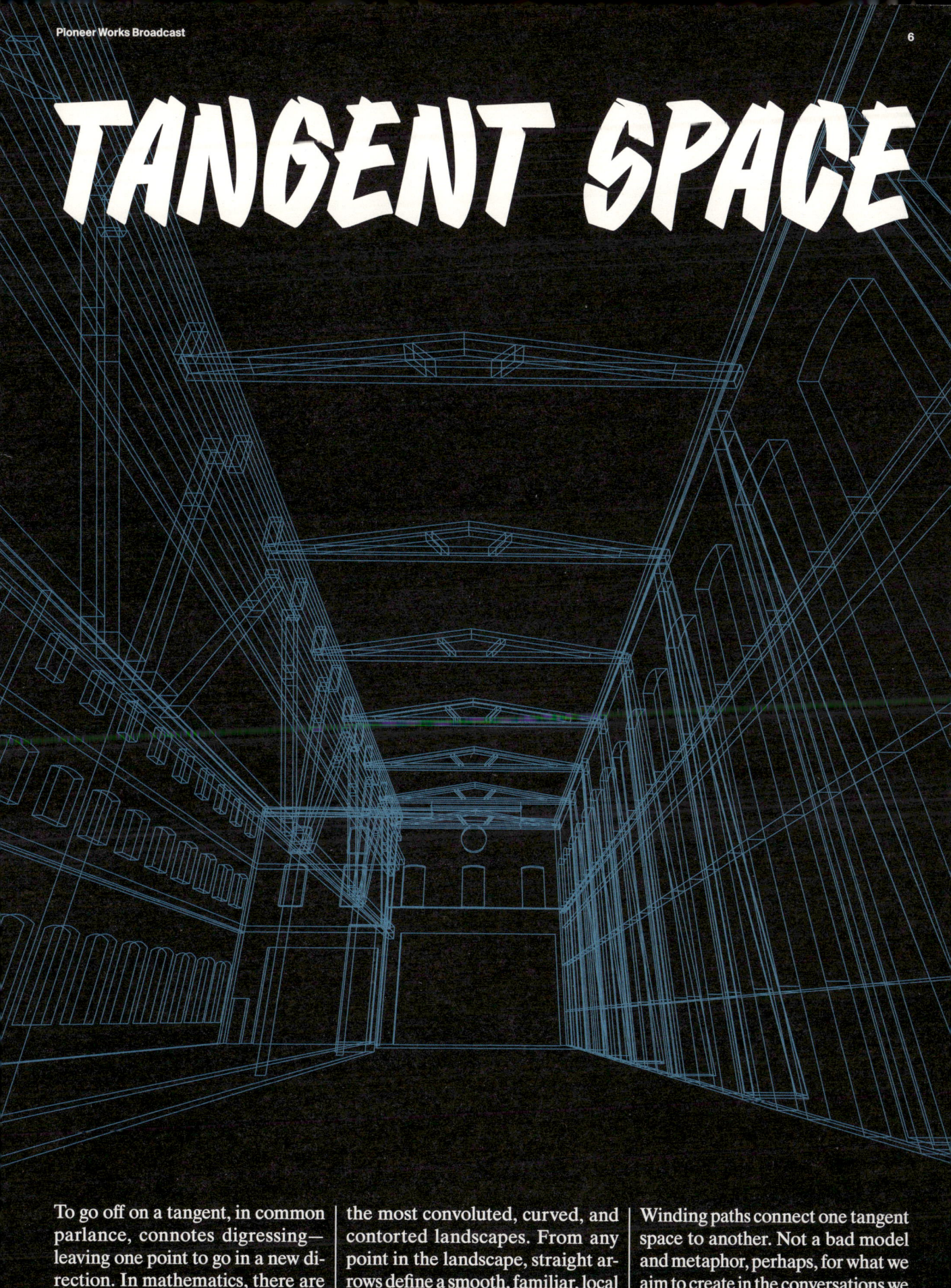

To go off on a tangent, in common parlance, connotes digressing— leaving one point to go in a new direction. In mathematics, there are not only tangents, there are tangent spaces. They help us make sense of the most convoluted, curved, and contorted landscapes. From any point in the landscape, straight arrows define a smooth, familiar, local map of up and down, left and right, forward and back: a tangent space. Winding paths connect one tangent space to another. Not a bad model and metaphor, perhaps, for what we aim to create in the conversations we curate at Pioneer Works—and seek to foster with *Broadcast*.

1
"'Are we done with psychoanalysis?' is a really important psychoanalytic question."
Judith Butler

4
"The best scientists I've had in my lab are the ones who are very 'chimpy.'"
Frans de Waal

9
"I would rather die than go to an analyst. If you light every corner of a house brightly, it becomes uninhabitable."
Werner Herzog

21
"Even though my work's funny and absurd, if it's not somewhat vulnerable, it's a joke."
Dynasty Handbag

10
"Sontag saw these women—these feminists in particular—as anti-intellectual, and that was the big sin."
Sigrid Nunez

15
"I wanted to create a space that felt simultaneously incredibly safe and incredibly dangerous."
Kristin Hayter, aka Lingua Ignota

3
"Sometimes you don't know how traumatized people are until you sleep with them."
Bessel van der Kolk

2
"Hysteria is a state that's responding in a very sane way to insane circumstances."
Aimee Meredith Cox

13
"There's no such thing as a perfect circle in physical reality. A circle exists only in our minds."
Janna Levin

19
"We don't see with our eyes. We don't hear with our ears. We hear and see with our brain."
Charles Atlas

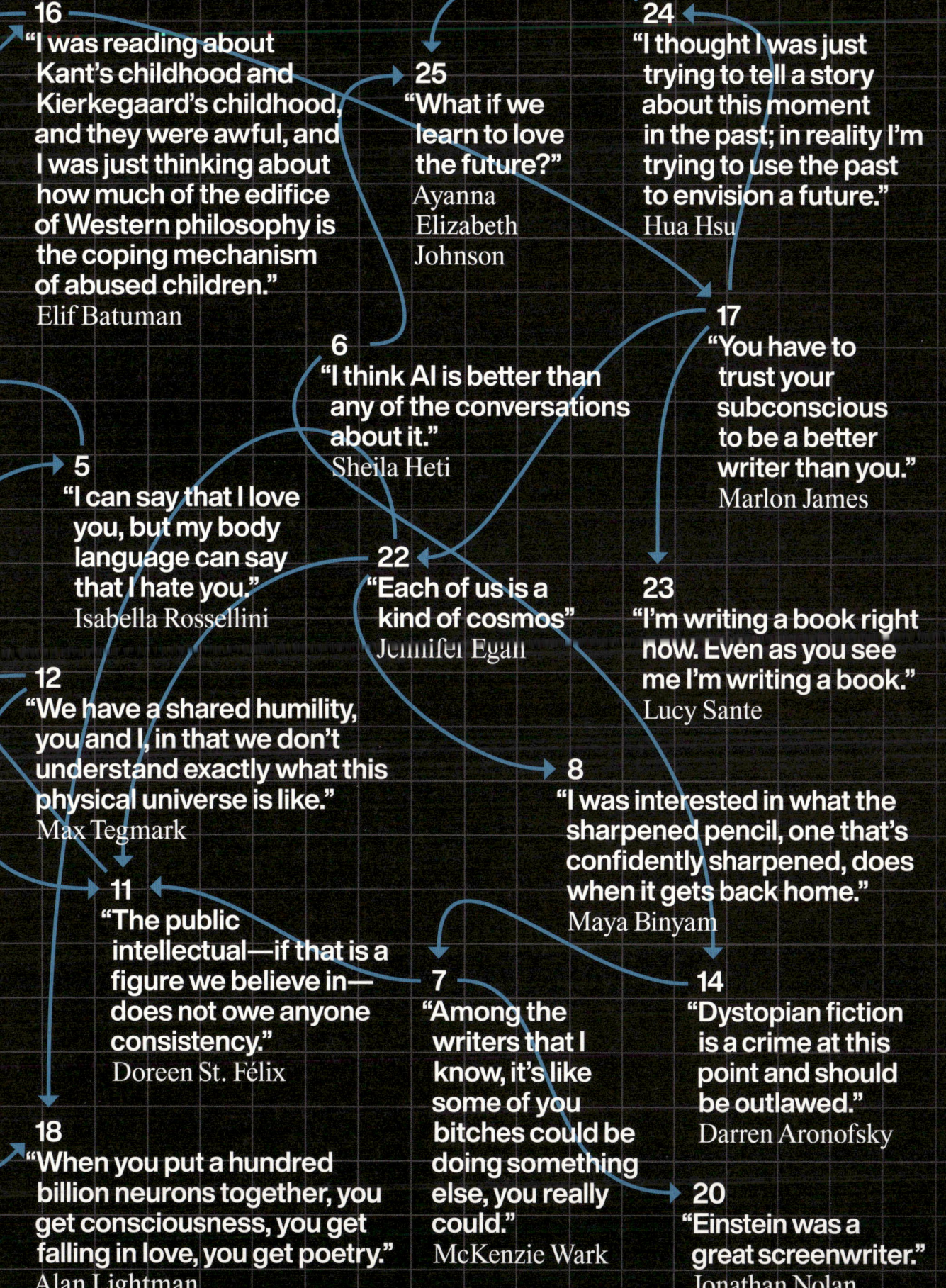

16
"I was reading about Kant's childhood and Kierkegaard's childhood, and they were awful, and I was just thinking about how much of the edifice of Western philosophy is the coping mechanism of abused children."
Elif Batuman

25
"What if we learn to love the future?"
Ayanna Elizabeth Johnson

24
"I thought I was just trying to tell a story about this moment in the past; in reality I'm trying to use the past to envision a future."
Hua Hsu

6
"I think AI is better than any of the conversations about it."
Sheila Heti

17
"You have to trust your subconscious to be a better writer than you."
Marlon James

5
"I can say that I love you, but my body language can say that I hate you."
Isabella Rossellini

22
"Each of us is a kind of cosmos"
Jennifer Egan

23
"I'm writing a book right now. Even as you see me I'm writing a book."
Lucy Sante

12
"We have a shared humility, you and I, in that we don't understand exactly what this physical universe is like."
Max Tegmark

8
"I was interested in what the sharpened pencil, one that's confidently sharpened, does when it gets back home."
Maya Binyam

11
"The public intellectual—if that is a figure we believe in—does not owe anyone consistency."
Doreen St. Félix

7
"Among the writers that I know, it's like some of you bitches could be doing something else, you really could."
McKenzie Wark

14
"Dystopian fiction is a crime at this point and should be outlawed."
Darren Aronofsky

18
"When you put a hundred billion neurons together, you get consciousness, you get falling in love, you get poetry."
Alan Lightman

20
"Einstein was a great screenwriter."
Jonathan Nolan

THE

Most of the world's fungi
remains mysterious to us.

WORLD'S

We should get to know
them while we can.

SECRET

BRAD BOLMAN

FABRIC

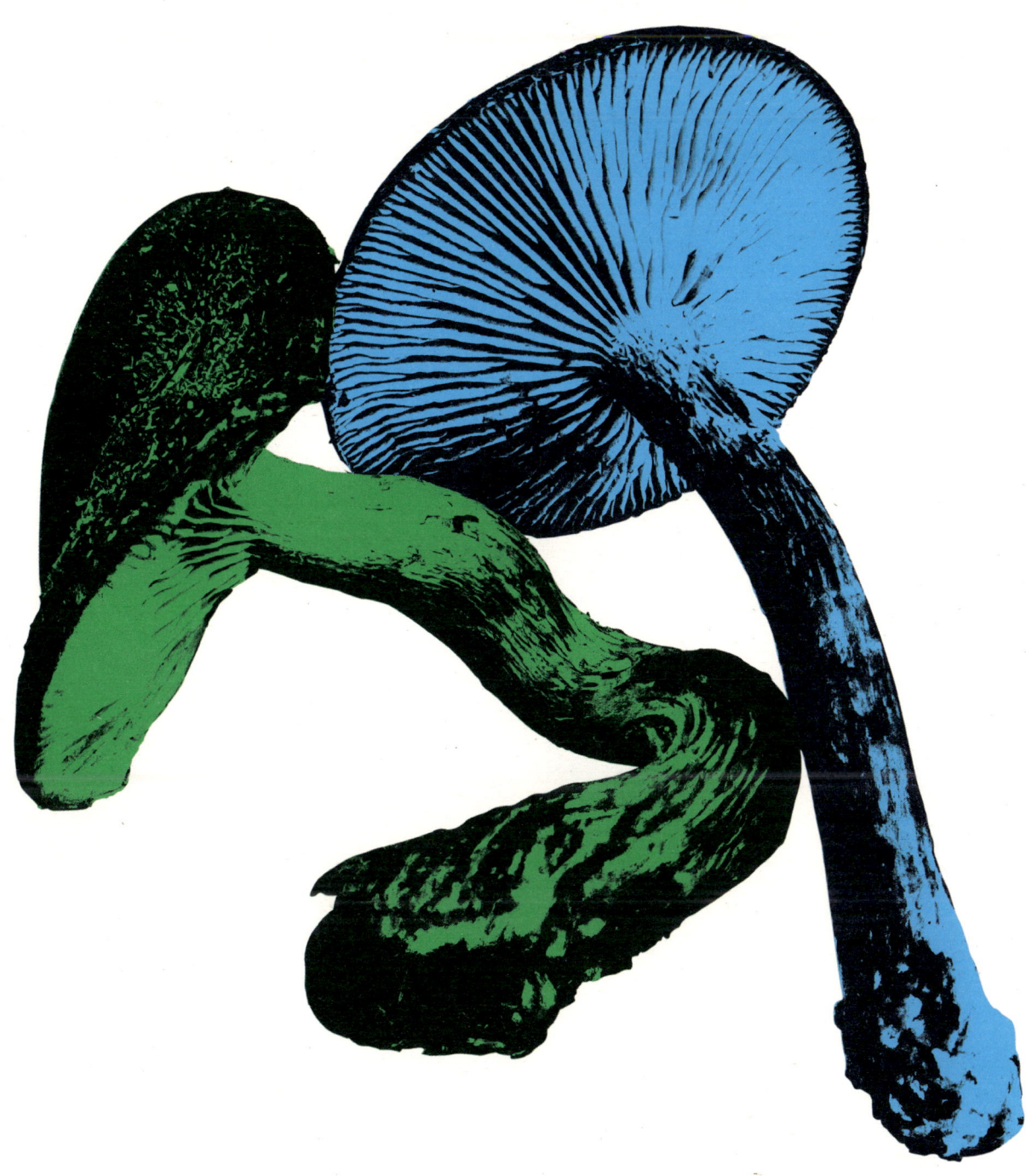

There are fungi in our oceans and our prestige television shows, in our pharmaceuticals and designer furniture, in our newspapers and our tumors. Wherever we look these days we seem to find more. It's like the morning after fall rain on a global scale. Yet, according to a recent estimate from the Royal Botanic Gardens in England, most of the world's fungal population—and, by extension, most of what we mean when we talk about fungi—remains mysterious to us.

A vast majority of the world's fungal species have, so far, eluded conventional scientific identification. Some of these species can be discerned by the presence of their DNA in soil samples, but we don't really know what they are. These unidentified life-forms, what researchers sometimes call "dark fungi," might be thought of as the world's secret fabric: the dark matter of life. They hold together our ecosystems in ways we are only starting to fathom. And in this age of mass extinction, their elusiveness has become an ecological concern.

The question of how many fungal species exist in the world has vexed researchers for over a century. In 1902, the most comprehensive taxonomy of fungi then in existence, Pier Andrea Saccardo's multi-volume

Sylloge Fungorum, listed over 52,000 species. They hailed from around the globe, although identifications tended to be densest where mycologists, the students of fungi, were densest, too: in Europe (especially England, France, and Germany) and North America.

There was reason to believe, at the time, that the number was inflated. Despite Saccardo's best efforts, there were duplicates and misidentifications—the result of researchers working in different places and languages, overlooking each other's writings, seeking to make a name for themselves by giving their names to fungi.

Saccardo believed his list was only a fraction of the global population, predicting the actual number of different fungi to be closer to 150,000. Today, after more than a century of additional work using high-tech genetic sampling techniques and global collection, researchers have cataloged almost exactly that many species. And still, there are strong indications that Saccardo's estimate was conservative. In 2011, mycologist Meredith Blackwell calculated that there might be nearly five million total species of fungi, and others have offered numbers as high as ten or twenty million, with more sober guesses around three million. We have found only a small fraction of the world's fungal diversity.

There are more fungi than plants and mammals combined, a diversity and extent matched only by insects and bacteria. A growing body of research has also revealed the essential role fungi play in maintaining ecosystems at nearly every level. They are decomposing chemical spills, turning ants into zombies, cycling nutrients, and weaving intricate webs of hyphal threads that transmit vital signals beneath the skin of the earth. Many of these species are threatened by a warming world, and we may not even know what they are or do.

Our uncertainty about their number raises deeper questions about

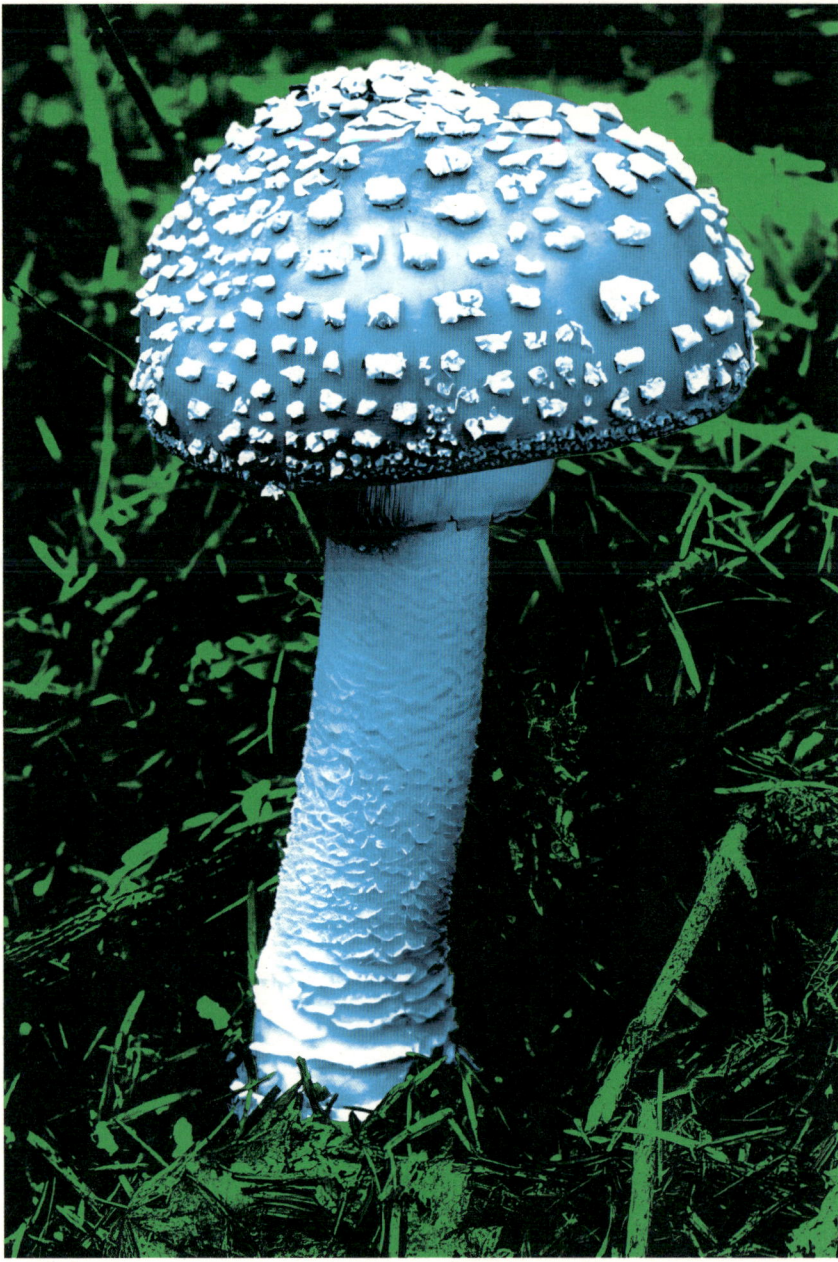

the curious nature of modern taxonomy. Indeed, it unsettles our very definition of a "species." Many unidentified fungi have in fact already been collected—they just cannot be visualized or cultured in laboratories. They are, in multiple senses, invisible, even as they represent a vast portion of our ecological tapestry. The quest to understand and catalog these "dark" taxa—living organisms that can't be brought to light—is a pressing task for science, and raises vital questions about the nature and value of scientific knowledge in our era of advancing technology and warming seas.

To get a sense of the fungi that's hiding in the shadows, one must first appreciate what fungi we can see—and how we got to know them. In the early nineteenth century, when mycology began to don the garb of a conventional scholarly discipline, the work of identifying new species focused on macrofungi. Brilliantly colored fly agarics and delicious boletes were hard to miss, if you knew where to look. Research relied heavily on experience and direct observation. Those interested in finding fungi—which were still considered "plants"—took extended strolls through quiet woods, picked scandalously phallic mushrooms, ate a few of the safer ones, and dried their most unusual spoils for later observation.

Unfortunately, in contrast to flowering plants, paper turned out to be a poor host for mushrooms: when pressed between sheets of parchment, they lost their lively colors and shriveled up. Drawing and painting were therefore essential, and many early fungal experts (or their wives and assistants) were remarkably capable artists, enabling colleagues in distant lands to witness their discoveries.

Few garnered envy quite like Charles and Louis René Tulasne, two French brothers who left law and medicine to dedicate themselves to mushrooms and lichen. Charles's careful drawings of life histories, symbioses, and microscopic structures are photorealistic and fantastically alive, dispatches from a vibrantly alien world. Mycology was at once an art and a science, and the enduring appeal of fungal illustrations, whether on notebooks or totes, confirms the lasting power of that amalgam. But by the end of the nineteenth century, the Tulasnes' drawings were replaced, in compendia of fungi, by lists, figures, and photographs. "Commercialism has killed the possibility" of breathtaking fungal illustrations, bemoaned mycologist Louis Krieger in 1922, "men are no longer training their minds, eyes, and hands for such work—the art is dead!"

If the art was dying, replaced partially by a desire to "let nature speak for itself" through photographs (what some authors call "mechanical objectivity"), mycologists were still being trained in new modes of seeing. The widespread accessibility of powerful microscopes took them closer to their objects of fascination than ever before. In his first book, Pier Andrea Saccardo described feeling "almost drowned" in a "great sea" of microscopic life-forms. He and

others like him could see what earlier eyes could not. The total number of identified species of fungi exploded.

Technology was one reason why. But Curtis Gates Lloyd, a key figure in the history of American mycology—he funneled part of a Cincinnati pharmaceutical fortune, at the turn of the twentieth century, into his passion for puffballs—identified another factor: the narcissism of other researchers. Foragers insisted on attaching their name to every little fungus they found, even when their distinctions between species were specious, a "fever" that Lloyd, in the early 1900s, named "species-making." Men looked at the world's fungal diversity, Lloyd thought, and mostly saw themselves.

However the acceleration of species identification occurred, Saccardo's near drowning indexed an irony: few mycologists were extending their search to the world's great seas. By 1950, dozens of studies revealed an immense array of aquatic fungi, many adapted to saltwater conditions. Identification required new approaches to an increasingly central mode of mycological inquiry, the so-called "pure culture," which allowed living organisms to be maintained over longer durations in specially fabricated containers. Culturing solved part of the preservation dilemma that plagued earlier generations—drawings are less critical if one can see the fungus itself—and revealed the unbelievable diversity of substrata on which fungi could survive. They existed on insect skeletons, onion skins, pollen grains, algae, shrimp skeletons, snake skin, eel worms, ferns, and almost everything in between. What did the recognition of these new homes for fungi mean for scientists?

For one, a billowing wave of species identifications and an explosion of fungal data that made life as difficult as it was exciting for working mycologists. The application of the electron microscope, beginning in the 1950s, allowed mycologists to

peer ever deeper into the microscopic world, illuminating the ultrastructure of hyphae, the tubular threads by which many fungi explore their environments. If visualization remained essential, here was a powerful new way to see. Yet with new tools came new problems: the known fungal world was growing, but so were the horizons of what mycologists didn't know. In 1962, British mycologist Geoffrey Clough Ainsworth worried that information about fungi was "increasing exponentially." In the 1980s, the introduction of DNA sequencing fundamentally reshaped how researchers name and classify fungi. It would also exacerbate, and soon validate, the concerns of researchers who had noticed major inconsistencies in how science had, prior to the advent of genetic analysis, arranged and classified the organisms they study.

Yeasts, rusts, and smuts, for example, are all fungi (and among the most alluringly named subgroups). But until surprisingly recently, yeasts were classified by their morphological characteristics and ability to assimilate carbon compounds, while rust and smut species were instead often named for the plants they parasitized. Thus brewer's yeast, *Saccharomyces cerevisiae*, literally the sugar fungus of beer, is named for its use of sugar as a carbon source, while *Ustilago maydis*, the smut of corn (or maize), is named for its plant host. Drawing taxonomic connections on the basis of DNA sequences seemed to offer a better solution, leaving each species division "a product of nature rather than the creation of the taxonomist," in the words of yeast taxonomist Cletus Kurtzman in 1985. Molecular objectivity seemed to offer a solution to the limits of mechanical guesswork.

The usage of genetic tools caused a radical redistricting of the fungal kingdom. Longstanding taxonomic groups fell apart, as things so often do, and phylogenetic studies drew new lines across traditional

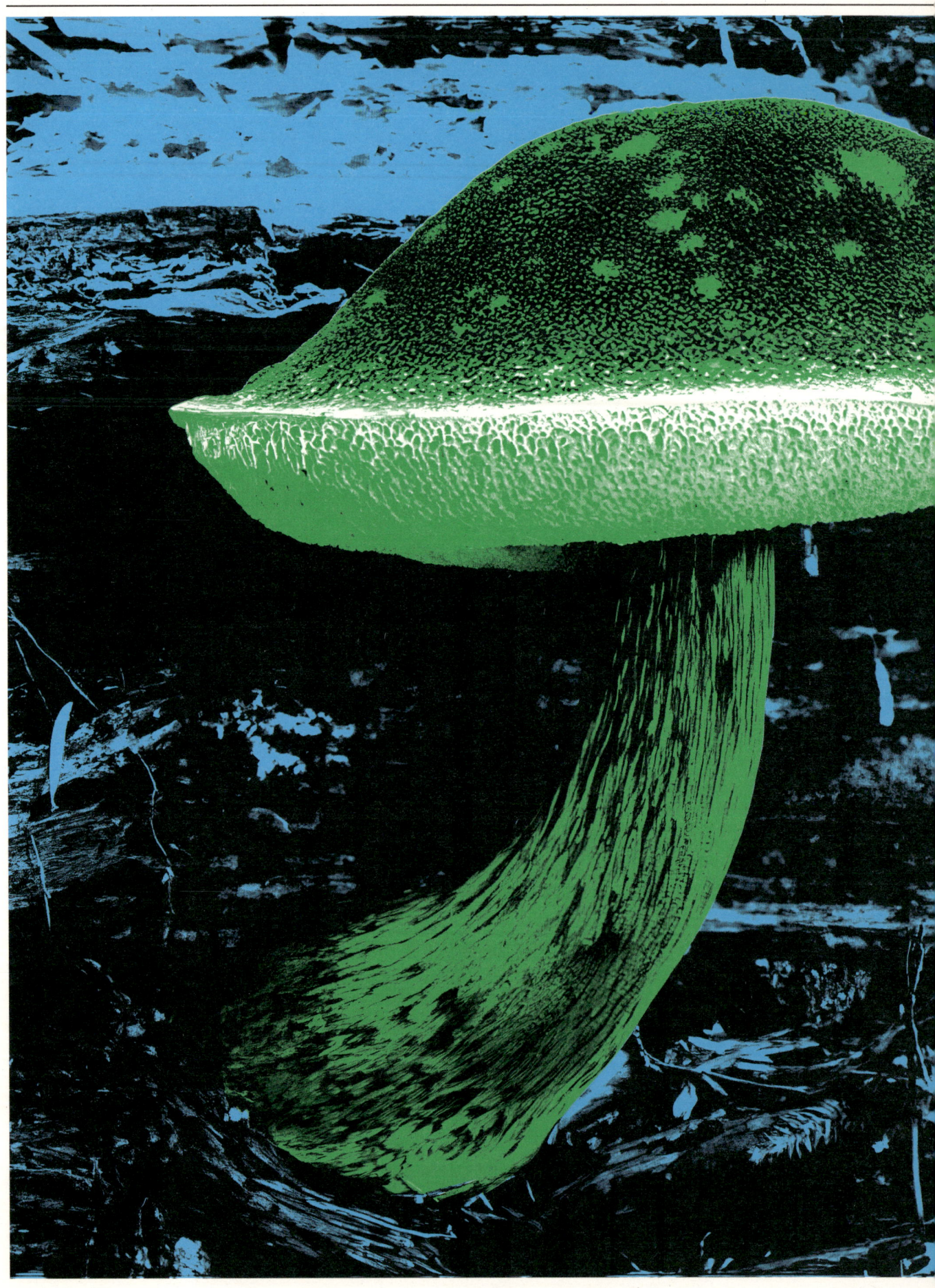

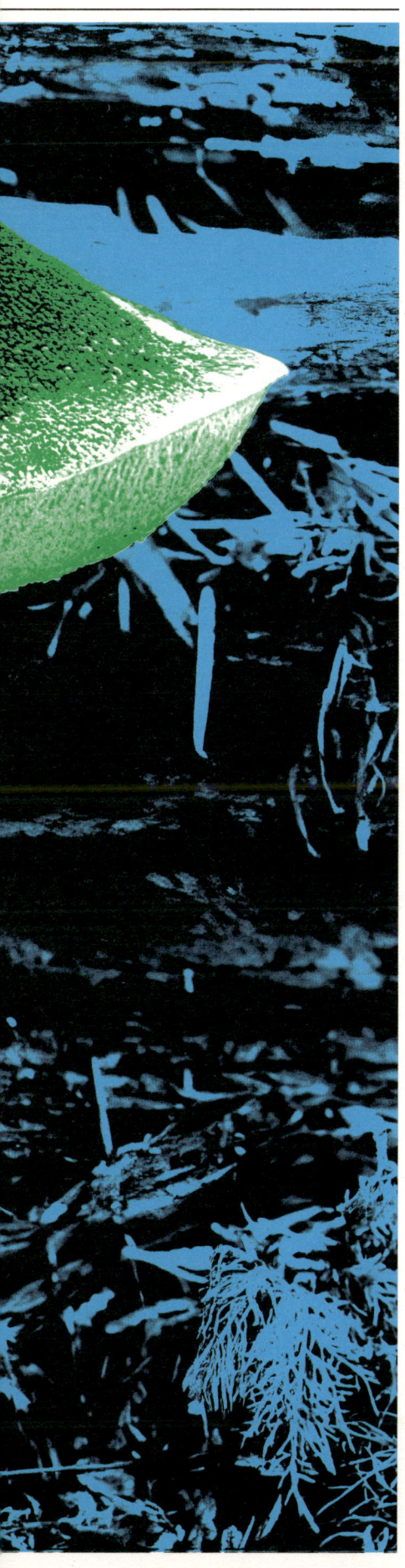

classifications. Many organisms were found to be, in fact, fungi, while others long studied as fungi, such as the late potato blight, were determined to be nothing of the sort. Some single species even turned out to be immensely multiple, like a lichenized fungus in the genus *Cora,* which DNA studies recently revealed to be several hundred distinct species. At the highest level, fungi, associated for centuries with plants rather than animals, were shown to share, to the mild discomfort of many vegetarians, far more with chickens than roses.

The application of advanced genetic methods to fungi was expensive and time-consuming at first. It could not be done willy-nilly on whatever potentially exciting speck of earth one found: the first fungal genome, that of brewer's yeast, was sequenced in 1996, but another important laboratory species, *Neurospora crassa,* followed only in 2003, and the results revealed major divergences in the genetic makeup of fungal families. Yet a decade later, researchers could economically collect large environmental samples, like tropical soils and vials of ocean water, gather DNA and RNA sequences, and find evidence of numerous previously unidentified organisms.

There was, however, a problem: many of these organisms were so small or fragmentary that they could neither be visualized by conventional devices nor cultured in laboratories. As mere read-outs of DNA printed on computer screens, they were difficult to interpret—the invisible obverse of the known and named world: dark taxa, or dark fungi. For mycologists, the existence of such cryptic, invisible, and "voucherless" organisms represents a multiform dilemma: a "staggering limitation" to communicating and interpreting fungal ecology and evolution, according to a recent study. The reasons are various, but they concern an ongoing reevaluation of how we decide what is and is not a species.

At a basic level, taxonomists name and classify organisms. The name *Candida albicans,* for example, coined by mycologist Christine Marie Berkhout, tells us the genus and species of the yeast commonly responsible for skin infections. Etymologically, it's a goofy name, since both *albicans* and *Candida* come from Latin words for "white," making *Candida albicans* something like "white whitening," the Wonder Bread of fungi. There are around two hundred other species within the genus *Candida,* including *C. antarctica* (found in a lake you'll never guess where, as well as on polished rice in Japan) and *C. krusei,* which limits the bitterness in chocolate.

Species and genus names frequently differ in reference, from places and characteristics to the discoverer's name, but the key point for working taxonomists is that each name has been published, usually in a scientific journal, along with a description. Under the terms set out by the International Code of Nomenclature for algae, fungi, and plants, each also requires a "type" specimen, a kind of standard representative, which might be a dried sample, an illustration, or a metabolically inactive culture. As Adam once gave names to the animals, so traditionally have mycologists given names to the fungi.

Dark fungi do not have names. For much of recent mycological history, the path to a name went via study of an organism's morphology and phylogenetic placement: how it behaves, develops, and forms relationships. But because dark fungi can neither be seen nor cultured, the task of conventional description is all but hopeless. Rather than isolated individual entities, they remain ghostly, fluorescently labeled presences in environmental samples.

One way of getting around this unnameability is DNA "barcoding," in which a small fragment of an organism's genome is used as a standard of comparison for its taxonomic group. (There is, to wit, a "Consortium for the Barcode of Life.") For fungi, such comparisons are usually based on small DNA regions known as internal transcribed spacer (ITS) groups. According to a 2018 study from mycologists Robert Lücking and David L. Hawksworth, the vast majority of the nearly one billion fungal ITS regions collected in a large database known as the Sequence Read Archive, operated by the National Library of Medicine, are known only by their sequence: tens or even hundreds of thousands of novel taxonomic groups. Naming fungi by their codes would eliminate some of the poetry of old names, but it could offer a way of getting a handle on these beings.

In doing so, however, barcoding threatens another death for the art of mycology. Some wonder whether there is danger in sequence-based naming. An emphasis on this practice could, for instance, diminish support for international culture collections, which often maintain a tenuous hold on institutional funding. A future sequence-based mycology might also stack the deck in favor of large laboratories with access to the most advanced sequencing tools.

There is debate about the validity of these criticisms, but few dispute the fundamental challenge that dark fungi represent to the work of mycology. "The fungal kingdom may be almost exclusively dark," noted a summary of a recent study in the journal *MycoKeys*. It no longer appears feasible, as it once did at the dawn of the DNA era, to apply a single, universal approach to identifying fungi. The future of mycology, and its implications for our understanding of life on earth, looks stranger than many could have imagined.

E ven assuming a limited estimate of 1.5 million species of fungi in existence, Lücking and Hawksworth argue that a complete inventory of them

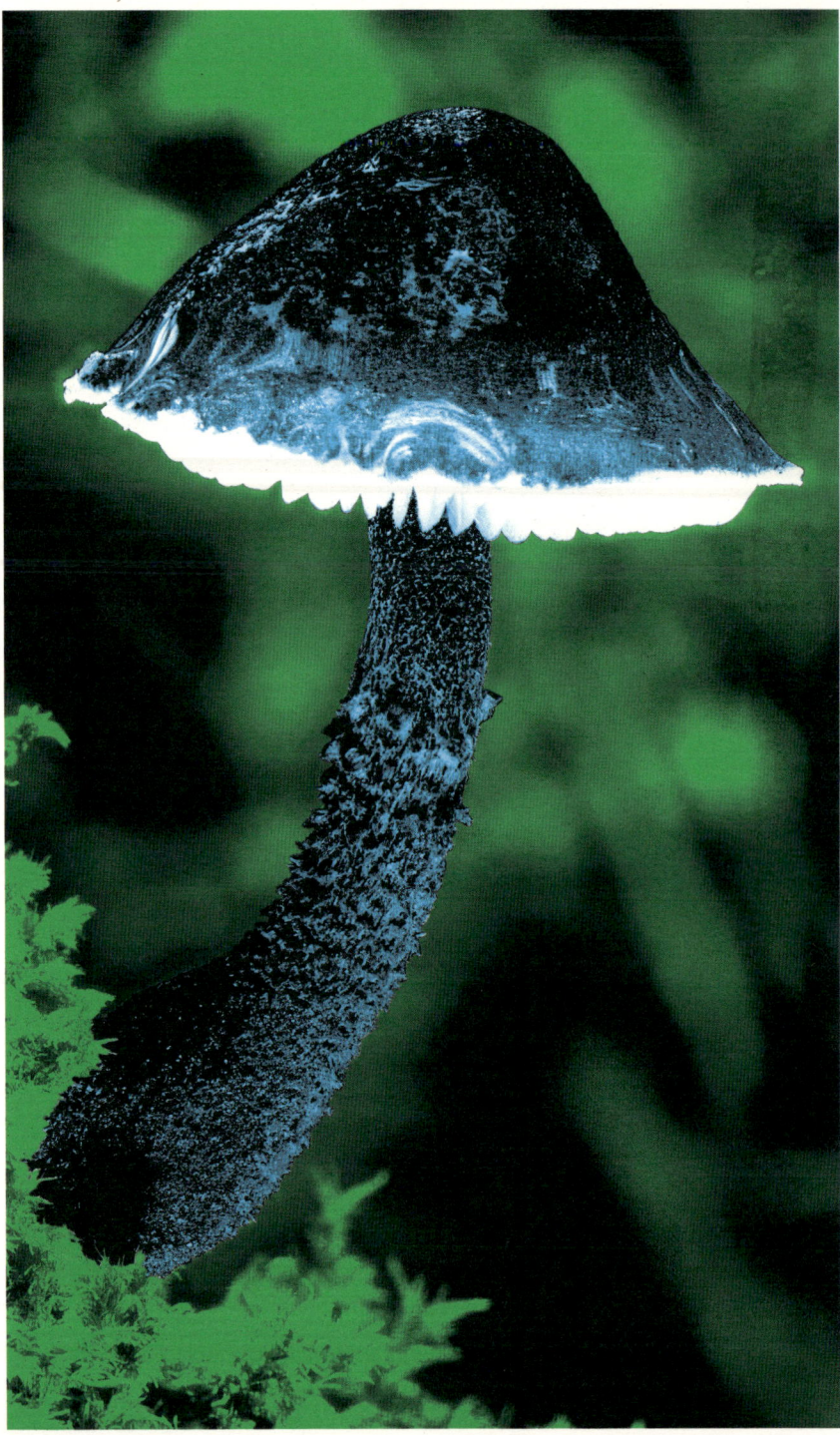

all would take 680 more years at our current rate of discovery. For their more expansive estimate of 3.8 million species, the full catalog would be finished only in the year 3848— and who can envision the challenges facing mycology, to say nothing of humankind, then?

"Imagine a census of a population of three million people," wrote Lücking and his colleague Conrad Schoch recently. "Now consider that only 150,000 of these people had a name and a valid, government-issued document connecting that name to the corresponding person. Imagine further that many of those people had more than one name, having been married one or several times or just because they felt like changing their name every once in a while. You end up with 150,000 people having 300,000 different names, while the remaining 2.85 million have none."

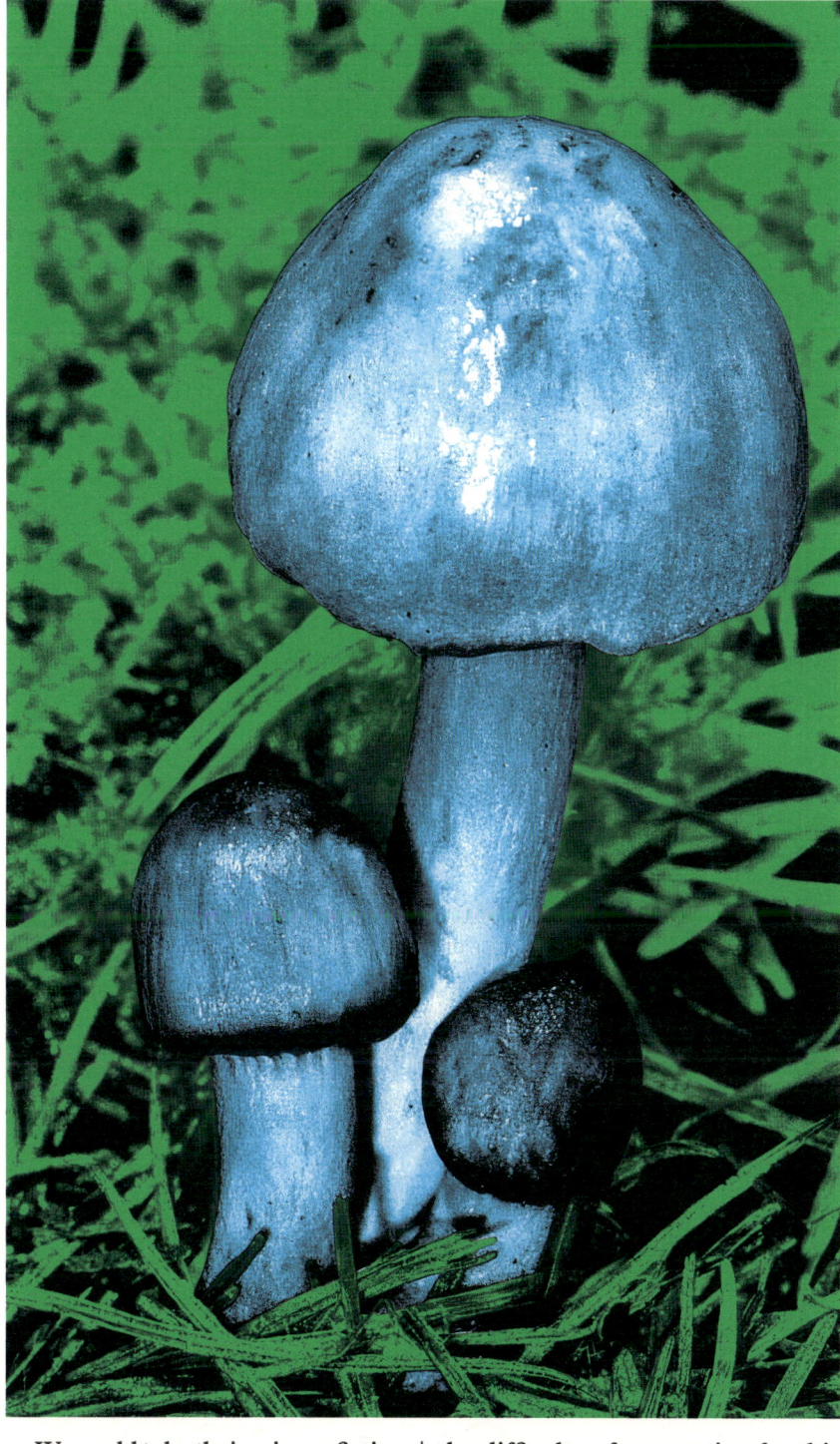

We could take their science fiction further: imagine that those 2.85 million people are dying, en masse and at an accelerated rate, leaving behind vital jobs that none of the survivors know how to perform. Many are unlikely to ever be identified before it's too late, and the plausible toll of such ongoing extinction is, in an era of climate catastrophe, monumental. Already vulnerable ecosystems may be further undermined, while

the difficulty of accounting for this vast world of dark life leaves many of our geo-ecological models fundamentally unstable. With so many medicines and industrial products reliant on fungi, there is good reason to worry that humans are destroying organisms with vital importance for the future of worldly life. How does a discipline with such a tight focus approach such a huge task?

In a 2011 lecture, "Dreams and

Nightmares of Neotropical Ascomycete Taxonomists," the Cornell scientist Richard Korf worried about the social utility of fungal taxonomy. Looking back over centuries of research, the famously eclectic Korf fretted that the old conditions enabling taxonomic inquiry, such as supportive museum posts and an academic environment of open-ended inquiry, were disappearing. The ivory tower, he noted, had ceased to offer such a climate, and funding agencies now held exaggerated sway over what studies could be done and how.

But Korf, who passed away in 2016, remained optimistic. Taxonomists seek the truth about relationships, he affirmed. Even more important than *naming*, which Korf had dedicated much of his life to, was the work of *saving*. Mycologists were in the business of preserving that which was being lost far too quickly, of "saving whatever we can of our natural diversity." This was, Korf concluded, in what he called his "final sermon," mycology's continuing moral duty.

What we colloquially call scientific progress is, in reality, a seesawing expansion of both the known and the unknown. A good experiment, after all, typically raises as many new questions as it answers. Although taxonomic debates, often derided as the quibbles of biological stamp collectors, have long seemed like footnotes to more thrilling tales of scientific discovery, they have a vital role to play in grappling with climate destruction: indexing loss and identifying what might be preserved.

If the work of naming and organizing species was once a mark of human mastery over the natural world, it might be seen as something else today, in an era in which that mastery finds itself so clearly upended: a form of attention, care, and consideration. Making sense of the darkness has its challenges and perils, but if we don't start looking closely now, we may only see some of the light after it has already gone. ●

HILTON ALS

The mighty critic and curator on Prince's eyes, becoming one's subjects, and writing free.

JOSHUA JELLY-SCHAPIRO

MY PIN-UP

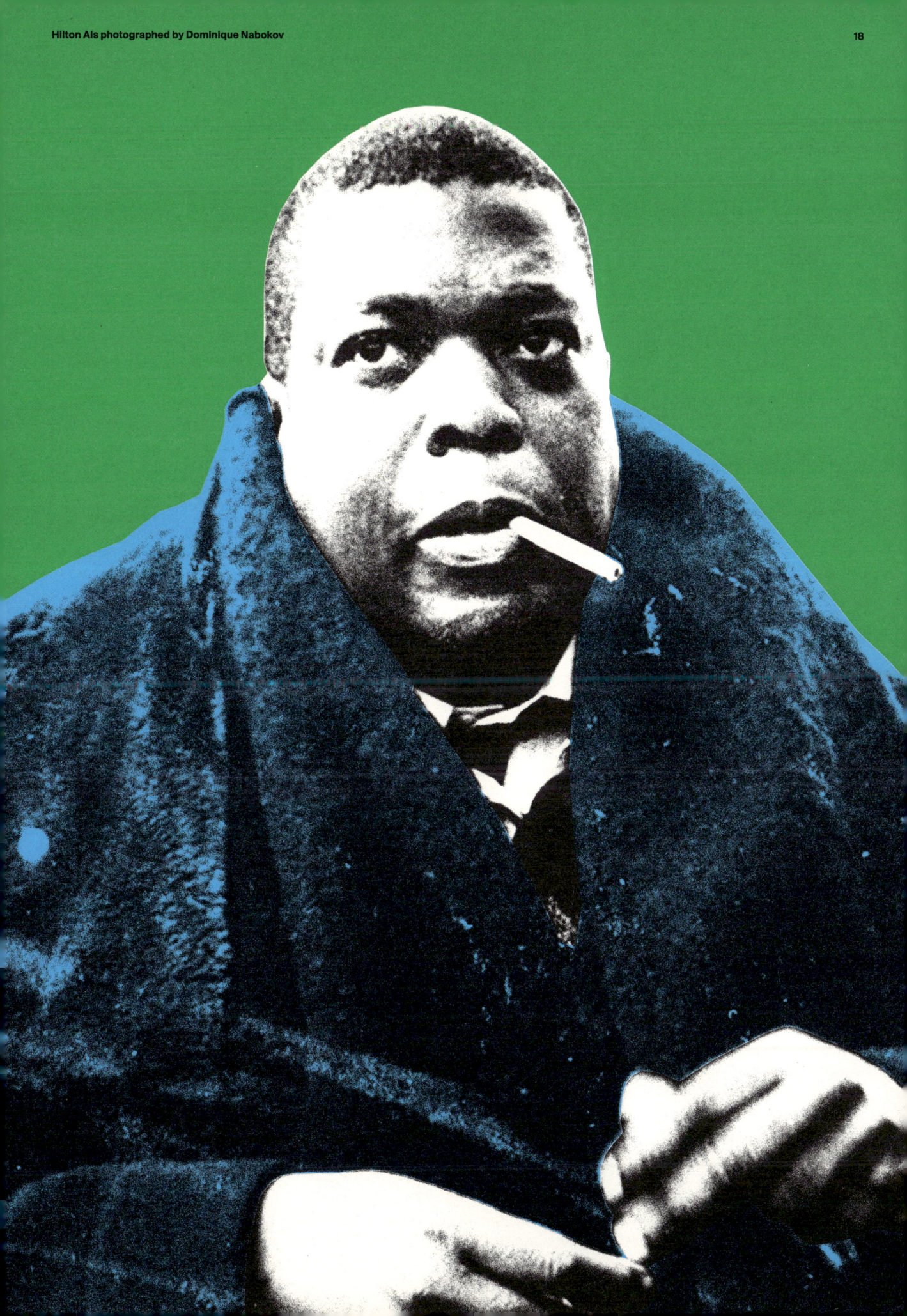

The essay, for Hilton Als, is much more than a form: it's an ethos for being in the world that animates not just his celebrated writing—his indispensable decades of contributions to *The New Yorker;* his cherished books—but every other facet, too, of a wide-ranging artistic practice that's as much about images and intuition as his ever-trenchant and revelatory prose. Hilton's Instagram feed lends the same sharply elegant language to lived experience that his criticism gives to painters and plays. As a curator, he's mounted exhibitions that figure the visual lives of writers who've nourished him—Toni Morrison, Joan Didion, James Baldwin—to craft "visual essays" as he calls these shows, "made not from sentences or clauses, but objects." He's a singular participant-observer in what we once called the culture, whose will to think in public has crossed and colored the linked worlds of art and literature, photography and fashion, theater and film. But Hilton's sensibility has always been about being present, in the most human and vulnerable ways, to those worlds that have shaped his own.

In this vein, Hilton is not the only American to have found his brave heart and dirty mind molded, from the 1980s to today, by the iconic songs and personality of the artist born Prince Rogers Nelson. But he is the only one to have turned an encounter with the late, great purple genius of music into a book-length essay, years later, that turns the conventional profile-form inside out. "Looking into Prince's eyes," writes Hilton in *My Pinup*, "must be like looking at the world. Or more specifically, the world of one Black man loving another." That world, and Prince's eyes, are but two of the subjects he explores in this essay on the artist, a portrait that becomes an oblique inventory of the writer's own self—and of the culture that made them both, in all its racial, sexual, and emotive contours and contradictions.

It was an honor to host Hilton's sole public event for *My Pinup* at Pioneer Works, in the borough where he grew up.

JOSHUA JELLY-SCHAPIRO
We met years ago in Berkeley—or *Berserk*-ley, as you call it.

HILTON ALS
Do you guys want to hear the story?

AUDIENCE Yes!

HA I was invited by the University of California, Berkeley, to talk to graduate students in English and Geography, which I didn't know was a subject. A very close friend of mine–my friend Valda, who I wrote about in *White Girls*—had just passed. I was in this beautiful place, and there was this twinkling little student in the first row, and that was Josh.

JJS Twinkling! Ha.

HA Twinkling. I was very sad and very by myself. Then there was an email from this person named Joshua Jelly-Schapiro, and I could not get over his name. That was the first thing. Then the second thing was that he said, "Do you have any friends here? Would you like to have dinner sometime?" He made himself available to someone he didn't know, who was passing through but who he had an intuition about. The point of the story is that we're here at Pioneer Works because of Josh's openness. You should take every chance you can in this life to be close to people who want to be close to you.

JJS Well, that is something that you've taught me so much about, the necessity and the challenge, really, as a writer, of being open and vulnerable to the world. We're here to talk about Prince. And about your beautiful habit of turning your engagement with art and artists you care about into writing, into art of your own. But I wanted to ask you about the first artist I know you were drawn to, and sought out, at a young age. You grew up in Brooklyn, and you admired the novelist Paule Marshall, who's from Barbados, as are

your forebears. You looked her up in the phone book.

HA Yes.

JJS Tell the story. How did your mother react to you doing that?

HA I adored my mother. She was a great reader. She had six children. There were four older sisters, and then there was me, my brother, and my sister closest to me in age at home. My mother loved Paule's book, *Brown Girl, Brownstones.* I remember picking it up and loving it, and wondering why my mother connected to this story of a first generation Bajan woman in Brooklyn. I thought Paule would be excited to meet me. Why wouldn't she? My mother loved me.

JJS Why not?

HA I looked her up in the phone book, and there was her number. She lived at 101st Street on Central Park West. I called her up and told her how much I loved her book. I was 10. She was like, "Really?" I just went on, and I told her about my mother, and about our life in Brooklyn and Bajan families, and so on. I said, "And I would love to come see you sometime." She was like, "Oh, well, okay." When I told my mother what I did, she thought I was dreaming. She looked and saw the phone book and that I had circled "Marshall, P." She just sat down. She couldn't believe it. I think a week later, I got on the subway and figured out how to get uptown, and I went to Paule's building, and her mother came to the door and wouldn't open it. It was something that I was very familiar with: a West Indian woman of a certain age. She said, "She's not home." "Well," I replied, "will you just tell her that Hilton came by?" You know, like a suitor. Then I went home and I told my mother. She sat down again. It has always been my habit that if you love something and your mother loves you, why wouldn't you want to visit?

JJS That's continued to be your habit as a writer and a thinker—the will and willingness you've had, which is sometimes fraught, to engage people

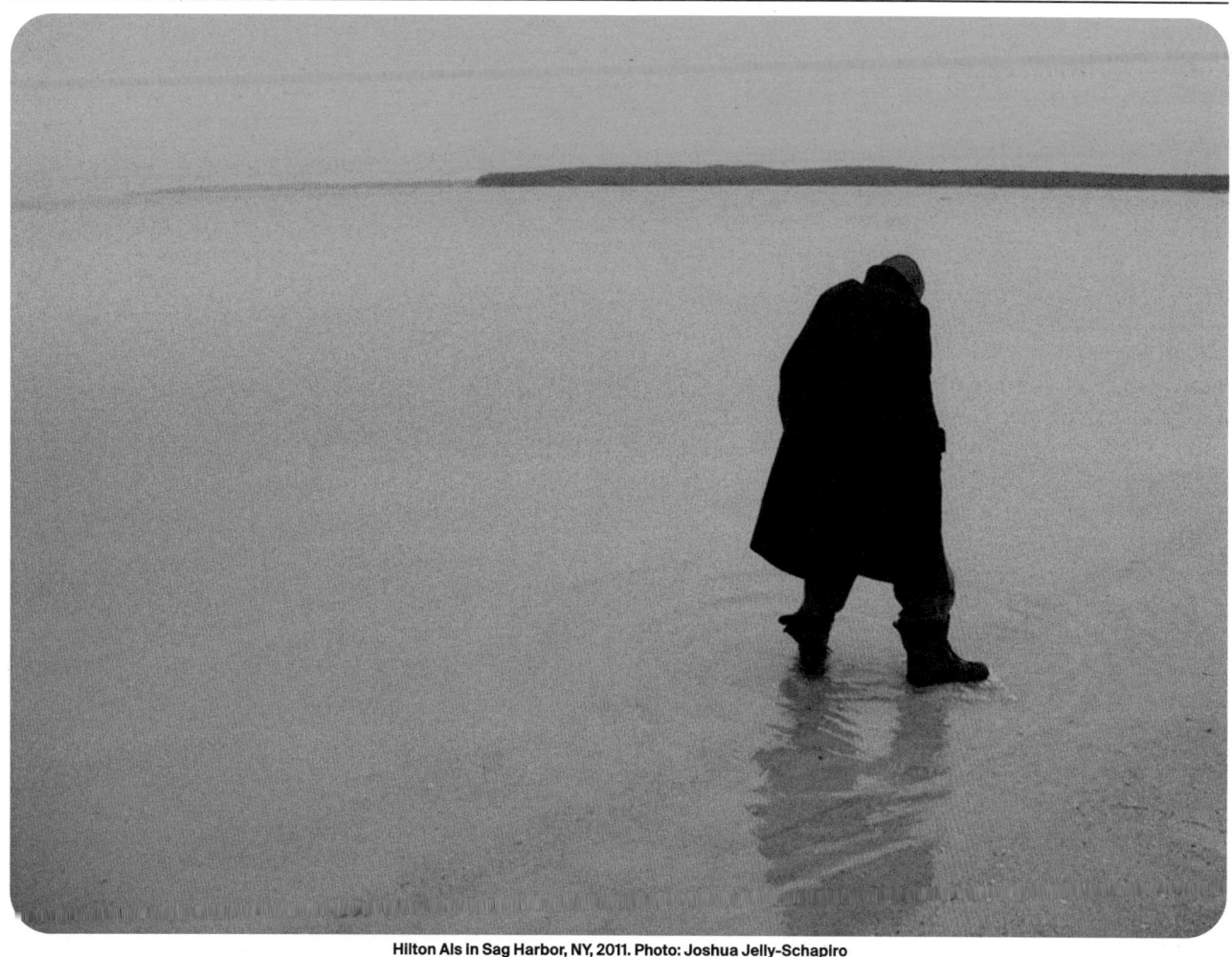

Hilton Als in Sag Harbor, NY, 2011. Photo: Joshua Jelly-Schapiro

whose work you resonate with on a human level. For much of your career, that's something you've done in the context of magazines, in the idiom of the magazine profile. In this new book, as in other recent projects, you've turned the profile form on its head, pointing to its inadequacies as a form. Long before you wrote about Prince, you were drawn to his records—to those first albums he released in 1978, 1979, 1980. *For You*, and then the self-titled one, and then *Dirty Mind*. Bring us to when you first discovered his music, to when you came to value him as a performer and an artist.

HA For many years I lived in Crown Heights and Bed-Stuy, and there was a record store named Sam Goody's on Nostrand and Fulton, somewhere around there. It would often have very interesting windows. One day, there was a picture of what I thought was a man, with a mustache and a blowout—a combination that sort of terrified me in the way that Little Richard terrified me, right? Because I was like them inside, and nobody was supposed to know that I was like them inside. He was 19, and then maybe a year or two later he was on *Saturday Night Live* in garters and a raincoat with his band. Watching him and no longer being afraid—because he was also making music—I remember this feeling of complete liberation. His early songs were so incredible to me because they were not really so much about objects as much as they were about identification.

JJS There's a beautiful line in the book where you say that when Prince hit with those early records, in particular, it was the first time a Black pop star was not limited by Blackness. What you meant by that, I hasten to add, was not something essential or essentializing about Blackness; you're talking about the ways in which Blackness is performed and enacted in music.

HA It's important to remember that I grew up in New York. In the city, no one was just one thing. For example, I was madly in love with a girl whose mother was Puerto Rican and her father was Jewish. We would have Passover, and her mother would make pork.

JJS A New York Passover.

HA My experience of New York was always that it was an amalgamation of cultures and people, and [Prince's] arrival was a breath of fresh air, because it was representative of worlds that we understood. I remember feeling this enormous relief in his artistic presence. There were other people that I would discover subsequent to Prince who were raising these questions, but in terms of pop music and a pop sensibility he wasn't frightened the way that Michael Jackson was frightened. "Wanna Be Startin' Somethin'" is a masterpiece, because it's about Michael's anxiety. He says, "They eat off of you. You're a vegetable," and he's speaking in the second or third person about himself, whereas Prince would often talk in the first person. But he would also bring in a character that he identified with, whether it was Dorothy Parker or his cousin or whomever. There was always an inclusion of a self that wasn't supposed to be you, but was you.

JJS In that sense, he's an essayist, right? Someone who brings in different stories, different characters. Obviously, his songwriting is extraordinary and he's a virtuosic musician. We know these things, but one other aspect you write about in the book—and particularly when he comes out and, *Wow, what is this? What is he not afraid of?*—is that his bands modeled a kind of female empowerment that wasn't about suffering, but joy. A genuine queerness that was about being open to who you are, how you want to be. In the worlds that you came of age within, people referred to Prince with the pronoun *she*, right? Michael Jackson as well.

HA Yes. When I go back now and think about music simultaneous to Prince, the B-52s were also super important to me, because again, they were transgressing what a white band was supposed to do, which was to *not* make danceable music. But then here they were, some white people from the South, who had the locution of Black people that I knew in New York, and at the same time they had great rhythm, and they could dance.

I remember they were on the cover of a magazine—*Melody Maker* or something—when they first came out, and the tagline was: "The B-52s Say They Don't Hate White People." Then you open the article, and they are all white. At CBGB's, in New York, there was a lot of confusion around them, because it was

It Hurts At First

Whenever I think I have become too smart
for something, it roars back. To outgrow,
in early thinking, meant to outstrip or exceed,
to leave behind, a liberty. Humans outgrow
(and so do sheep) the tyrannies of weather,
parents, some superego terror governing
from on high. A poet once said in a modern
declaration that what she loved when she
was young, what she would not disavow
in front of clergymen or enemy, what was
constant despite power or her own shyness,
she would always love. A brave-faced child
who insists the ugly is not ugly, who sees shame
as merely a place to park and not part
of the body. A prophet-child. Once, looking
for a sign from the universe, I saw Fran Lebowitz
on the subway and mouthed I love you to her
as I got off my stop and the doors closed.
I swear, she raised her brows and I giggled, girlish,
brazen with a small edict, barely audible,
facing almost no consequences at all. My loves
are private games until they are not games.
Until a spilling occurs. A thousand small storms
at the threshold of my skin. A seam emerges
and breaks. No seamstress rushes to mend.
There's nothing to mend. Out, I come.

—Megan Fernandes

dance music. I'm always very drawn to a confusing cultural element, so if it's a band of gay men and two female singers who speak Southern American in music, I'm there. With Prince I was also there, not because he was singing out of marginalization but because he was singing out of his own originality.

JJS That was particularly true with his first records. Then, in the '80s, he became interested in pop stardom and achieved it with these massive records—*Purple Rain*, "Little Red Corvette," and all that—that weren't as interesting to you. But he won you back, later on, with *Lovesexy*. You thought this was someone you could write about in a fruitful way. And you finally did so after you met him

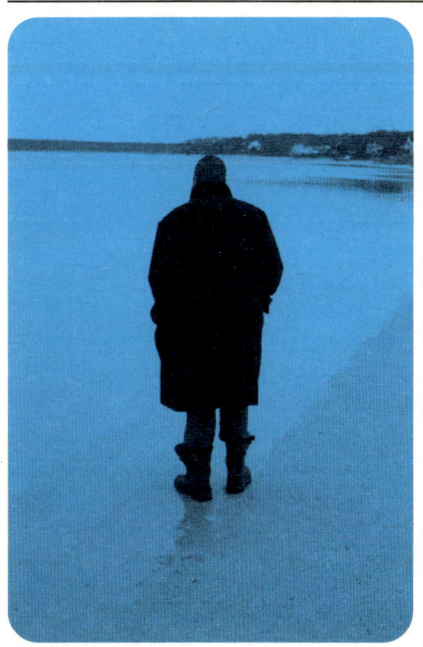

for the first time in 2004, during his tour for the *Musicology* album. You write about meeting him backstage, in St. Louis.

HA He was really beautiful. I mean, he was stunning. I just have to remind you of that. He looked like the most beautiful turtle I'd ever seen, because he had this long neck that came out of his sweater. He was tiny, and you just wanted to kiss him and help him with his homework. [laughs]

JJS You've talked before about falling in love with Andre Leon Talley, who you wrote your first *New Yorker* profile about. It's a process of falling in love and then extricating oneself—or not. It can be very tricky.

HA It is very tricky. You're in the process of loving them while you're writing about them. You *have* to love them, and then you have to go away. You have to be critical of the love that you feel. A funny thing happened recently. I'm writing a very long piece about Angela Davis. She's just a great human being, and incredibly vulnerable and true.

I was leaving San Francisco, and I had the files for my story at my feet in a bag. I looked up, and there was Angela in the lounge. My heart started to beat the way it does when someone you're in love with is in the room, but I also couldn't speak to

her, because I was rewriting her. I couldn't say, "Hey, Angela."

JJS She's become a character.

HA She's in my mind, and she's in my bones *as writing* now. She can't be that separate person.

JJS It becomes a profound collaboration, but also you need the space to write it yourself.

HA You have to. During the time that you're spending with that person, you're not there. There's a great quote from Diane Arbus, one of my favorite artists, where she says, "I never rearrange the subject. I rearrange myself to see the subject." That's what I do, so sometimes those people are shocked by the piece, because I've rearranged myself for a year to *be* them. Then, sometimes, if it's a shorter, more critical piece about their work, it's less entangled. There's less projection. You're listening in a different way, and you're less exhausted. It's an exhausting process.

JJS Indeed—this process of falling in love, but also the ways in which you can't be present for that while writing an essay. But in *My Pinup,* you write about meeting him after this arena show, the conversation you had with him.

HA What he did that truly lives in my heart—as if it was yesterday— was when he had this enormous show in St. Louis, and I had been backstage. As I was leaving, I heard, "Mister, mister," from one of his makeup people. She said, "Prince wants to see you. He would kill me if I couldn't find you." And so we went back to his trailer and he was in the makeup chair, with that beautiful head. He turned to me and said, "Hey." Then he turned to the makeup lady, and he told her, "Number 14," and it was his eyelashes. He had just given a show for three hours. After all of that noise and theatricality and giving, he was able to see himself and be present—with his eyelashes. It was so profound.

JJS Then he or his assistant said, "Oh, why don't you come back to Paisley Park [Prince's home and

studio], and we'll write a book together," and you rejected him.

HA Oh, that was going to be trouble. I would have never left. He was really beautiful.

JJS One of the wonderful things you do in this book, around the edges of this meeting, is to write about your desire for love, your desire for admiration, your desire for twinship. It's a theme that appears and reappears in your work—in *White Girls*, in your essays about literal twins, and in *My Pinup*, through Prince and his performing partners like Cat [Glover], and your own desire for a twin. In the realm of writing, do you find it true that the people you're drawn to reflect something in yourself that you identify with, like it's yourself but not yourself?

HA One of writing's great liberties— or liberations—is that you get to explore. It doesn't cost you anything; it's a pencil and some paper. You get to investigate aspects of yourself that sometimes someone else has articulated. Marianne Moore is one of my favorite poets, and she quoted others quite a bit. She said, "If a person has said the thing better, why wouldn't you quote them?" For me, quotes that the subject gives are a way of understanding why I was there in the first place, that my intuition about the situation has everything to do with where I am at and where they're at. I almost never write about really famous people because they already have their face, or think they do.

Prince was an exception to that, in a way; the iconography is so tremendous that you can just get rid of it. I was taken by his small town manners, his understanding of other people, and his concern that you be entertained. That was his job. He wasn't smug and he wasn't superior about that work. He just didn't like record executives. But I think that I'm mostly interested in subjects that don't have their face yet. Even if they're putting on their number 14 eyelashes, they're still trying to determine who they are—he was that. We all are. ●

Close Encounters in the Fourth Dimension

REBECCA WRAGG SYKES

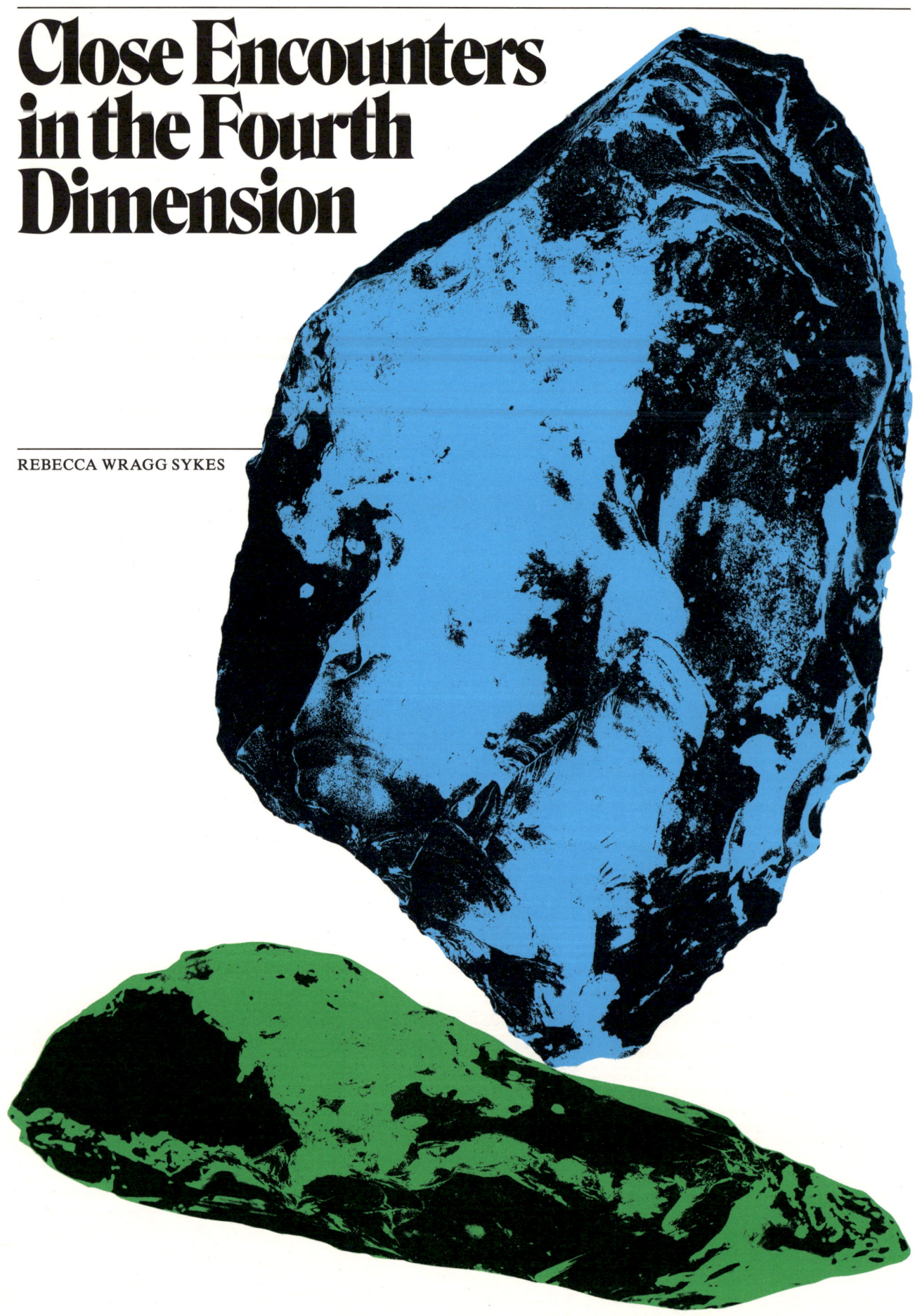

How rediscovering Neanderthals primed us
for the search for extraterrestrial life.

At some lost moment in the Pleistocene, over 200,000 years ago, early *Homo sapiens* wandered into western Eurasia for the first time. The world they found there was largely the realm of nonhuman animals. As they edged northwest into the great steppe-tundra, they came across species they had never seen in Africa. Plants were no longer grazed by gazelles but by reindeer; the bloodstained muzzles snarling at them no longer belonged to cheetahs, but to cave lions. Yet there was another, much more uncanny presence in these strange parts. Somewhere along their path, these pioneers would meet alternate versions of themselves: human, yes, but of another species altogether. Soon they would get to know the Neanderthals, and this relationship would forever change the course of both species.

Maybe they first stumbled on their artifacts. Imagine it like this: a group of early *Homo sapiens* round a cliff and see a cave open up before them like a great dark eye. Inside, as their vision adjusts to the gloom, among moss and dust and owl pellets, they find stone tools. These objects are recognizable to them, yet nonetheless strange; made in ways that are not their own. They creep farther in, exploring tunnels and chambers. Deep in the guts of the mountain, the dancing light of their pine torches illuminates a glittering film of translucent calcite, revealing something else almost familiar. Bones, but not those of cave bear or hyaena. Instead, a skull with a humanlike face, though longer and with deeper eye sockets.

Of course, perhaps this first contact was truly face-to-face, eye-to-eye—staring right into that original uncanny valley.

What's incontrovertible at this point—though it was deemed near-science fiction not more than a decade ago—is that among the encounters that eventually took place in the flesh, some were of a decidedly carnal nature. Numerous studies of ancient DNA prove that Neanderthals and *Homo sapiens* were having sex with each other; indeed, that this happened quite a bit. The genetic inheritance of that interbreeding, which began over 200 millennia ago, survives today: up to 2% of all living people's genomes are Neanderthal.

The context of these relations is harder to discern. Were they carefree flings born of curiosity, long-term relationships, or were they chance assaults, part of a wider pattern of subjugation? Perhaps alternating between all three, this interbreeding persisted until around 40,000 years ago. And only a few centuries after it stopped, the Neanderthals vanished altogether, leaving behind only hybrid babies and their descendants.

And then we forgot all about them. It took until the nineteenth century, amidst seismic scientific, social and economic change, for European industrial-military infrastructure to blast Neanderthals back into the light. We rediscovered them piecemeal, puzzling out faces and minds from dry bones and cold stones, eventually recognizing them as another sort of human. Neanderthals would become a foil for our desires, dreams, and demons. They inspired new conceptions, expectations, and imaginings about humanity in a way that is strikingly similar to emerging ideas about alien life. This connection might at first seem like a stretch, but offers many instructive commonalities. The discovery of Neanderthals and the search for beings from other worlds each challenged parochial perspectives on time, language, technology and culture—the very tenets of what had long been believed to set humanity apart from the rest of life as we know it.

I mmediately following their rediscovery over 160 years ago, Neanderthals started exerting a magnetic pull on human thought, whether scientific, literary or artistic, eliciting a cascade of emotional and cultural responses. They inspired wonder—quite literally, in the only recorded response to a Neanderthal skull from Charles Darwin—and a desire to connect across tens and hundreds of millennia. But they also caused deep anxiety, because of what happened to them: extinction. The realization of the Neanderthals' existence challenged our central metaphysical place in the cosmos. It drastically stretched our frame of reference, much like the development of "deep sky" astronomy.

By the early eighteenth century, astronomers had already absorbed the fact that the Sun, rather than an

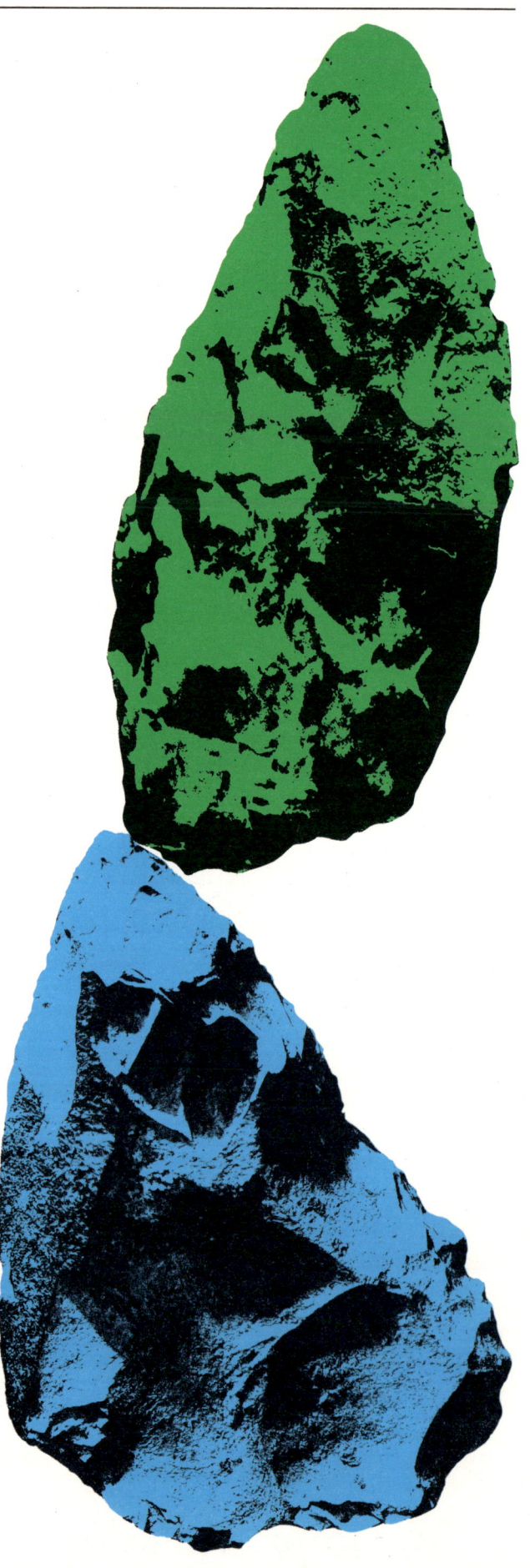

exceptional entity, was "just" the closest star to Earth. Around a century later, when the largest-yet telescope focused on fuzzy clouds beyond the visible stars, another alarming scale-shift occurred: farther off than the constellations of our own Milky Way, the heavens held yet more galaxies, downy with stars. And so, four decades after the first Neanderthal discovery in 1856, people were fizzing with ideas not only about life on the moon and Mars, but on planets circling other stars. Those speculations, both serious and fantasy, have since continued to build on each other. As we began to grasp the dizzying diversity of organisms on our own heavenly sphere, the staggering vista of potential life on worlds with different conditions of gravity, chemistry, pressure and temperature rapidly became vertiginous with possibility.

Researchers have since come up with a number of ways to detect simple alien life based on biosignatures—essentially, observable signs like telltale chemicals that point to living creatures processing energy or reproducing. Finding smart aliens might in fact be easier, because if they have a material culture, then we enter the realm of potential technosignatures: the manifestation of technology via perceptible, intentional manipulations of matter and energy. This could be in the form of radiation unexplainable via known astral processes, or tangible objects. The largest telescope feasibly within current technological capability could exploit how gravity of the Sun can bend space itself, forming a gigantic magnifying lens, potentially letting us see the surface of an exoplanet, and any megastructures built by forms of extraterrestrial intelligence, or ETI.

Searching for and analyzing bio- and techno-signatures is, of course, precisely how archaeologists like myself investigate the ancient human past, though we are separated from our subjects not by space, but time: the fourth dimension. We might excavate the ashy layers of a 100,000-year-old hearth, measure its temperature, analyze the fuels burnt, but we cannot *feel* the heat of its flames on our faces. In this way, archaeologists, astrobiologists and ETI researchers are all fundamentally trying to reach beyond the here and now: to map, describe, reconstruct and, in a real sense, *experience* other worlds of the fourth dimension, and the beings who lived there.

Thinking about how to detect technosignatures, whether across interstellar distances or in archaeological sites on Earth, brings us back to first principles, and the question: what is technology, and *why* is it? Technology allows two things: ever-more powerful and complicated ways to alter materials, and a means to store information. Ultimately, one key aspect of intelligence is about the *intentional* manipulation of the surrounding world or environment, in a way that goes beyond immediate survival. It's the manifestation of some level of consciousness. Archaeologists study the emergence of technology, but while we may draw on theories of

anthropology based on our own species, with Neanderthals we can't assume we're dealing with evidence of minds all too similar to ours.

Thinking in this way is proving useful for SETI (Search for Extraterrestrial Intelligence), too. And so, in 2018, I was invited to take part in workshops with the Breakthrough Listen project, with anthropologists and ETI experts thrashing out ideas about what technosignatures mean, the way in which they might appear, and how humans might deal with a situation where we actually make contact. Even in the absence of some sort of direct encounter, technosignatures can still tell us about more than just their makers' cleverness. Artifacts are the embodiment of ideas. A silver fork implies solid foods, a particular notion of eating and a value system including rare, processed metals; an alien signal implies a transmitter, implies a desire to communicate, implies an understanding of relationality between entities. The challenge will be making sense of these discoveries.

One of the most common questions people have about Neanderthals is whether they made art. Creativity remains one of the essential benchmarks for how we think about ourselves as humans—and in this way, technology and artifacts have implications far beyond mere functionality. Answering the question about art is always tricky because it's necessary first to parse what exactly we mean by "art." Almost all the time, the person asking is thinking of *visual* art, and often something akin to the layered lions and horses heads painted on the walls of Chauvet cave, or the carved female figurines scattered across western Eurasia.

No objects like this have ever been excavated from any Neanderthal site, but there is evidence of an aesthetic engagement with materials. Neanderthals altered surfaces by incising them, and used colorful substances like mineral pigment. They left behind other more ambitious projects, like the giant rings of stalagmite fragments discovered in a deep cave in France. This site, called Bruniquel, is so primal and weirdly beautiful it could come from any era. In its simplicity, it's also something we might imagine stumbling upon in a cave on Mars. In this case, the pattern-recognition system represented by our eyes and brain would be the technosignature-detection software. Thinking through the ways we identify aesthetic material engagements in Neanderthals and other hominins therefore has real significance for how we deal with the idea of an ETI signal that contains scant obvious symbolic content, or even none that we can recognize.

The notion of contact is based on the assumption that communication is possible. Humans communicate with each other using a multiplicity of media: by sound or movement, light or texture. What these all share, however, are underlying commonly understood meanings, whether expressed with a character that is easily trace-able to what is being referenced, or via symbols that require decoding. While it's been possible to discover, investigate and sometimes decode ancient communication systems like Mayan hieroglyphs, the farther back we go in time, the more difficult it is to reconstruct or even assess the presence of language. That's because, aside from the added complication of script changes, sound in human speech alters far faster than meaning. (Sometimes you can trace the mutations. For example, the English word *brother* and the French *frère* derive from the Sanskrit *bhrātr* and the Latin *frāter,* respectively.)

Even the most ancient potential shared word roots only go back a few thousand years. Deeper into prehistory, beyond any written texts, we face a linguistic abyss. This gives us some idea of the conundrum we might face with an extraterrestrial signal, where there is no contextual data at all from which to reconstruct meaning. The question of whether Neanderthals talked is therefore of relevance for approaching SETI. Anatomically, there's very good evidence that Neanderthals were both capable of making sounds a lot like ours, and also that their hearing was tuned into the same frequencies of speech. However, precisely how they spoke and *what* they spoke about remains unknowable.

If we do ever detect a technosignature, it is entirely possible that its makers could be long dead, or that the culture, species or even total planetary biota from which it sprang are extinct. Theoretical considerations of SETI require estimating the longevity of technology and cultures; we are certainly talking about timescales covering at least thousands, if not millions, of years (also taking lightyears into account, the space separating a signal from its source). While the origin of the signals might be technically "derelict," ruined traces of the transmission hardware might survive, and offer a material context for investigating the signallers themselves. Here, too, there are archaeological correlations: we can only see the barest outlines of an open-air Neanderthal campsite at the site of La Folie, France, marked by a circle of postholes enclosing scattered debris inside. But sometimes, when excavated, the edges of their stone tools are still sharp enough to cut a human hand.

Whether we are talking about intelligence in Neanderthals or ETI, we tend to focus on things like processing or memory in a computing sense, the ability to manipulate matter in complicated ways, and to connect ideas and make leaps of inference. Less attention is paid to emotional intelligence as a metric of adaptive importance. Humans have immense capacity for empathy on an individual basis and can demonstrate extraordinary levels of compassion and altruism at societal levels. Yet as a species, though we are exceptionally good at creatively using materials, extracting energy, distributing labor, and manufacturing a profusion of choice and stuff, we aren't

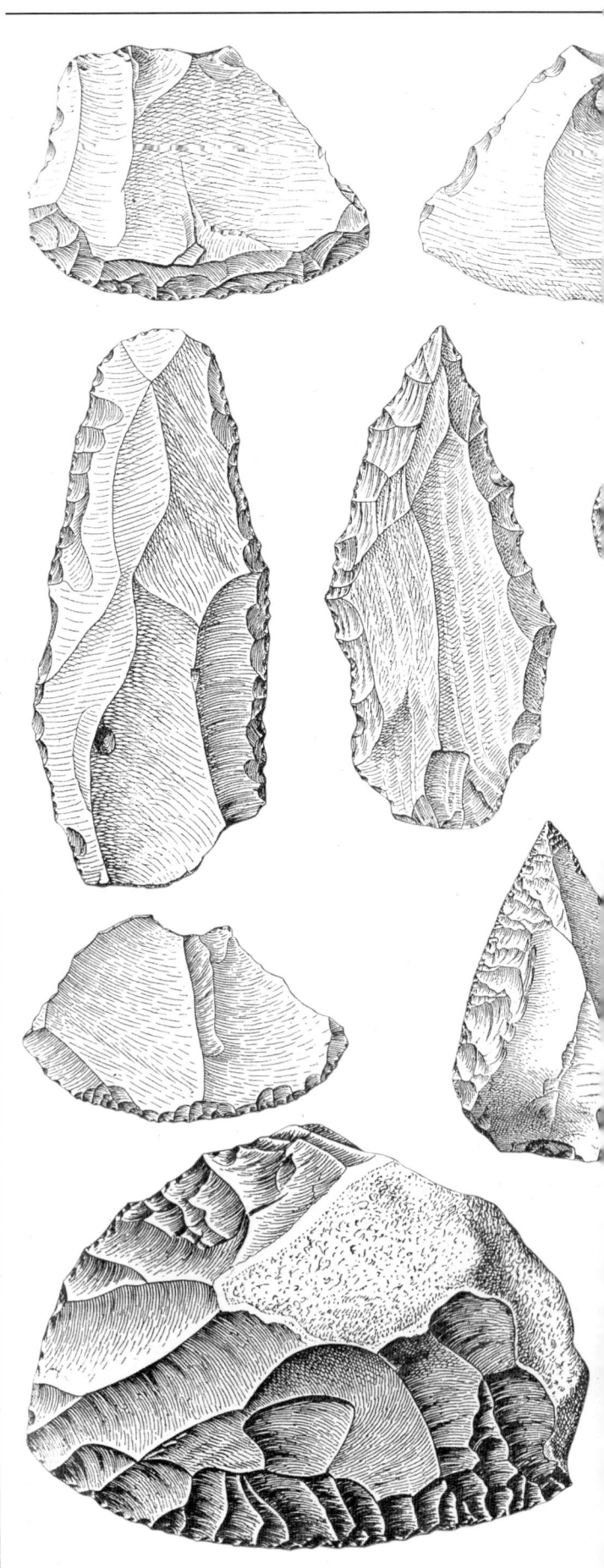

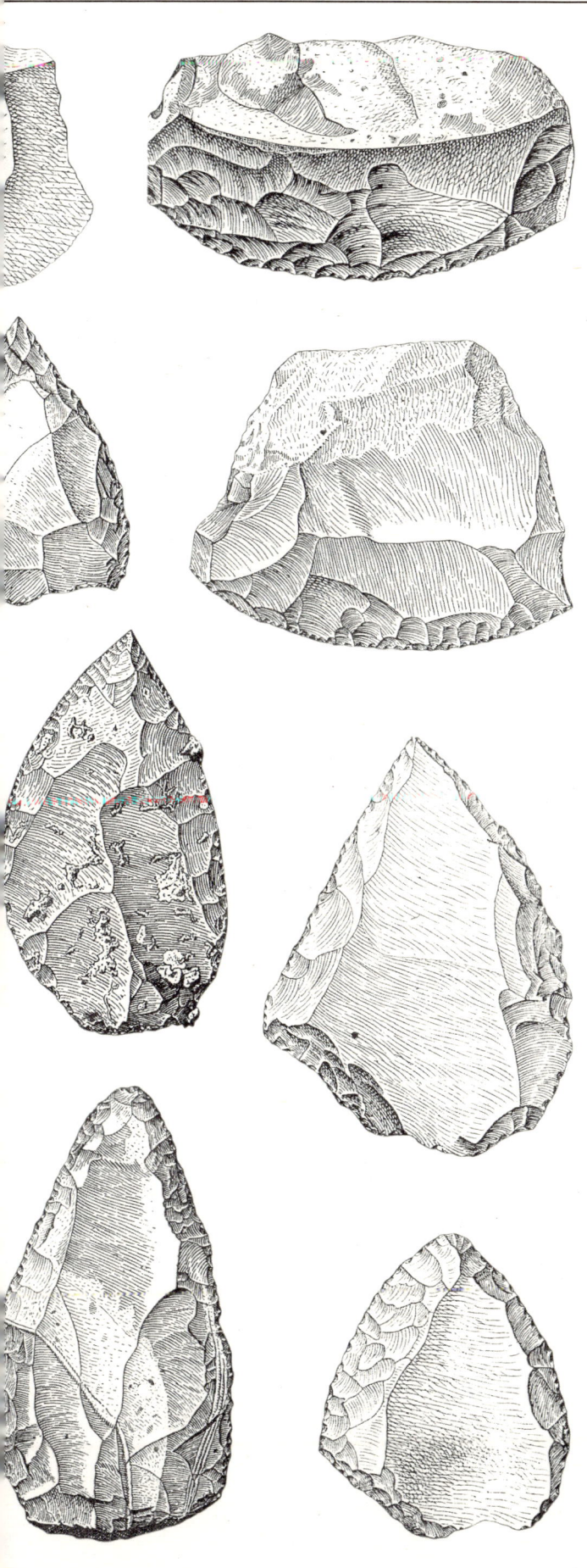

so great at doing all this in a truly equitable or sustainable manner. Most strikingly, notions of competition and exploitation for resources as the basis for "civilization" end up being projected onto other species, both nonhuman and ETI, as the definition of evolutionary success.

For example, discussion about Neanderthal subsistence has often involved comparing it to that of *Homo sapiens*, in order to look for evidence of dietary underachievement. This assumes that the Neanderthals' disappearance was caused, or hastened in some way, by their own incompetence. Archaeological discoveries several decades ago undermined once-popular theories that Neanderthals were scavengers rather than top hunters. Those ideas were soon replaced by suggestions that they were incapable of fully exploiting their environments by eating seafood, small game or plants. Today, we know that they were capable of all of this, and on occasion even appear to have over-indulged in some easily-available foods like tortoise. Nevertheless, the sense that being a successful hominin means capitalizing on resources rather than using them "lightly" and sustainably has persisted. This idea, that ultra-exploitation is a natural outgrowth of evolved intelligences, is echoed in some early technosignature theories of how ETI harvest energy.

Proposed in the mid-1960s, the Kardashev scale presents quite mind-boggling definitions of alien technological development. It held that the first stage of this definition of "civilization" here on Earth, or Type I, would be achieved once we were using *all* available energy, from solar, tide, wind, geothermal, nuclear, fossil fuels, and anything else. Type III, the most advanced "civilization" predicted by this scale, envisions a culture that has managed to gather the energy from the majority of stars in its *entire* galaxy. This means enclosing celestial bodies in gigantic spheres, which would dramatically alter the way such galaxies appear to distant observers, making them glow in infrared. In theory, this should be a relatively straightforward way to work out whether vast, smart ETIs exist. Using astronomical data already collected, searches for signatures potentially suggestive of Type III engineering have been undertaken, but so far revealed very little in the way even of possibilities. A new study examining over 16,000 sources found just two galaxies with an unusual infrared emission worth investigating further.

It follows that these Type III cultures must be extremely rare. But why? In 1961, the likelihood of ETI existence was quantified by the Drake Equation: if the physics and chemistry which led to life on Earth are universal, then it seems likely that "they" are out there. This led to the Fermi Paradox: if ETI likely exists, why the lacuna in confirmed evidence? The timely explanation—based on our experiences of the nuclear age—was that perhaps massive technological development inevitably led civilizations to self-destruction. Another more

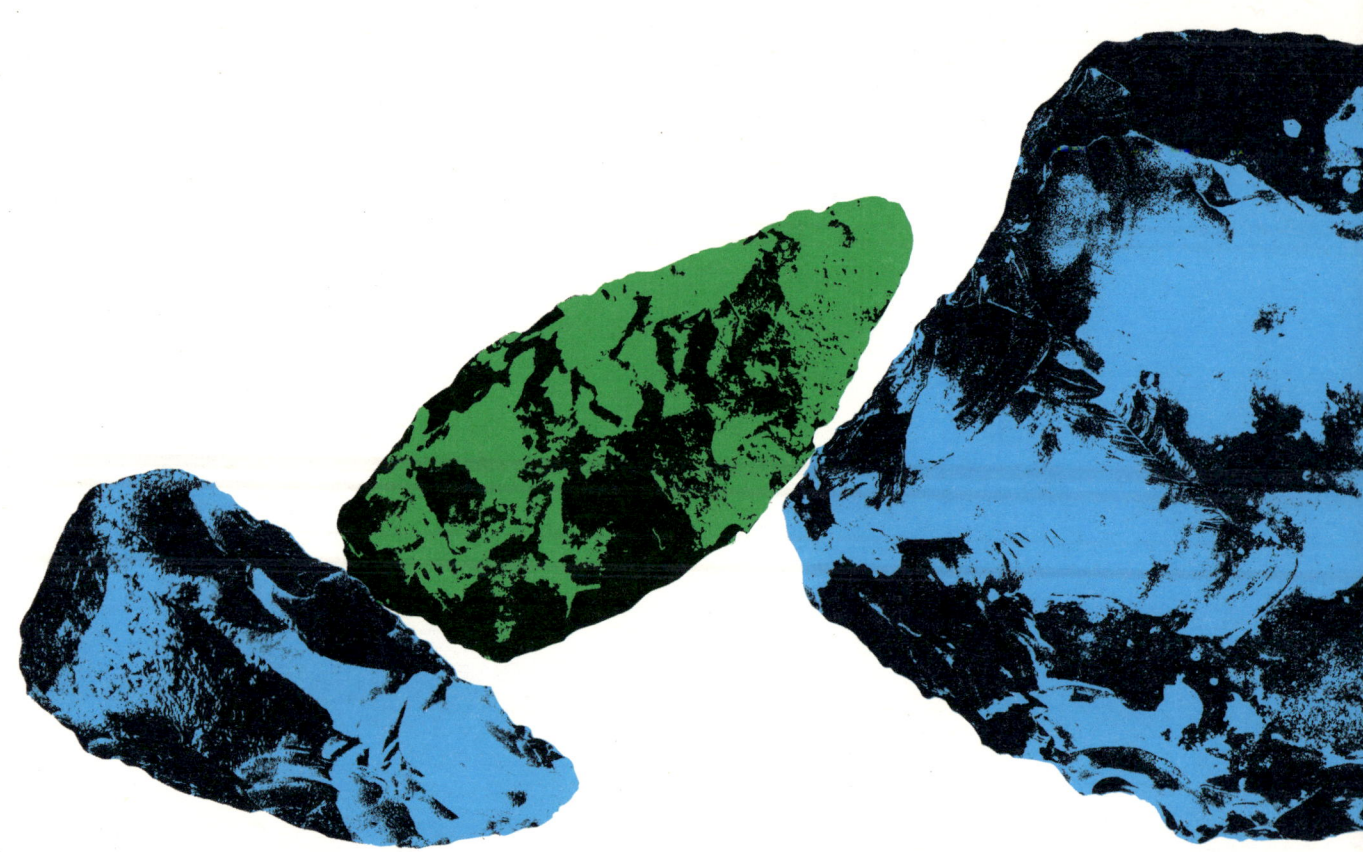

nuanced answer followed, based on something termed the Inevitable Expansion Fallacy. Somewhere along the line, it might become obvious to ETIs—as it is dawning on us—that rampant development is not the most adaptive choice, if it means maxing out every available resource at the cost of sustainability or quality of life. Instead, they might choose another path. In other words, the dominant, and decidedly capitalist, conceptualization of Earth's techno-civilization may be hampering our search for other lifeforms.

What if we did meet aliens? Science fiction often imagines such an encounter as innately hostile; either they mistreat us, or we mistreat them. In the latter scenario, they invariably fall victim to our scientific curiosity. We reduce them to specimens for testing, rather than an opportunity for new relations and exchange. There are echoes here too in debates over the future of Neanderthal research. Genetic engineering is allowing us to examine, through small agglomerations of cells, how Neanderthal-specific genes alter very early brain development and connectivity. The results are admittedly intriguing, showing clear differences, but there is an unaddressed ethical question. At what point do we stop such research? How large and diversified should we allow these *Neanderoid* cerebral blobs to get? What about connecting them to sensory inputs? Or splicing bits into living creatures; from mice to primates? What if some rogue lab attempts a

de-extinction, using a Neanderthal genome in an ape or human fetus, as is being pursued for mammoths? We should consider carefully where giving free rein to cold curiosity, rather than ethical empathy, could take us.

Thinking about the ways we approach studying both Neanderthals and any potential ETI from a moral and ethical point of view forces us to consider deeper questions about who we are as a species, and who we want to be. Pasts, present and futures are connected, and even though archaeology and astronomy more broadly lead us to confront the temporality of existence, our perspective on cosmological scales tends to be extremely myopic. We can just about comprehend centennial time, envisioning the handful of generations separating us from the Renaissance. Move beyond that and we struggle. Millennial timescales become fuzzy, while thinking across spans that saw the vanishing of entire other forms of humanity—Neanderthals, Denisovans, the Flores "Hobbits"— is solidly beyond our imagination.

The prospect of annihilation is especially difficult to process. The ever-present specter of extinction that hangs over Neanderthals is a reminder of the ephemerality of existence. The few hundred thousand years they were around is small beer within deep time paleontology, and even the much-discussed claim that *Homo sapiens'* presence and significant impact on Earth merits the naming of a new geological epoch—the Anthropocene—fades somewhat when looked at over properly ancient time. In fact, it was the *long durée* viewpoint of

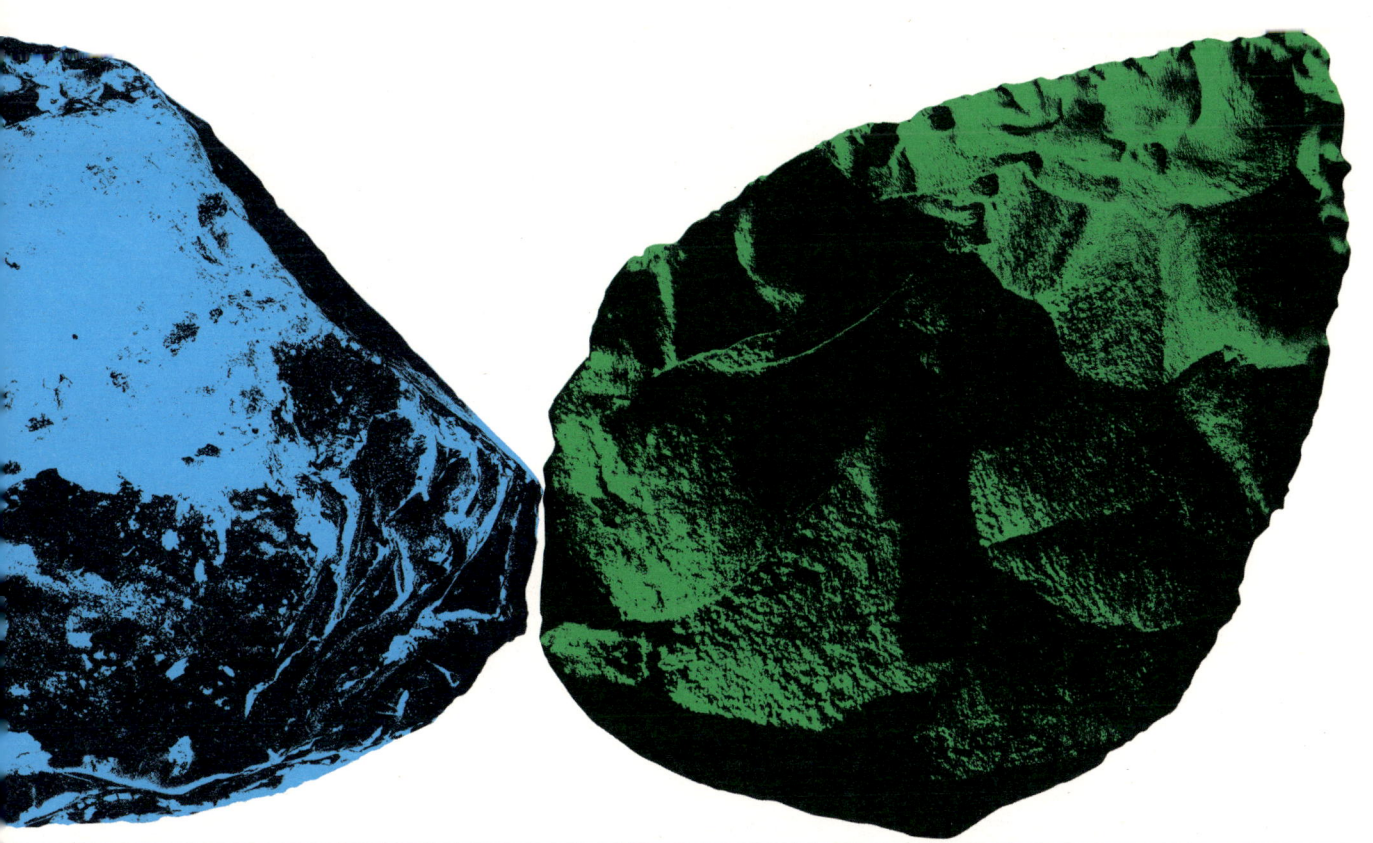

astrobiology that led some researchers to seriously consider the possibility that an intelligent species arose and developed an industrialized lifeway at some past point in Earth's history.

While the evidence was not convincing, the intellectual exercise did reveal something sobering. Even if some humans eventually leave this planet for other cosmic shores, what remains of us here when passed through the wringer of tens or even hundreds of millions of future years—and the inevitable continental shuffling and tectonic ructions—will be little more than geological perfume. The entirety of hominin material existence will on average be reduced to a rock layer of less than a hand's breadth. Aside from freakishly rare preservation of fossils or objects, our presence will only be traceable through relative amounts of chemical biomarkers and elemental isotopes, unusual molecular patterning, rare metal abundance, and possibly nanoscale plastics and synthetics.

One persistent notion found in science fiction consists of a "post-human" future, where various interfaces between consciousness and cyber data will allow us to transcend our "meat sack" embodied existence. In a real sense, Neanderthals have already ascended in this way. Rather than ceasing entirely to exist, their DNA still walks the Earth—in varying proportions and configurations—within you and me. When all that is left of a species is the blood running through the veins of another, is that extinction or survival?

Perhaps the greatest gift the Neanderthals bequeathed us—beyond genetic upgrades and perhaps even technological or cultural know-how—is the opportunity for self-reflection. What makes us different, and what do we share? In what ways do they influence what we mean by "human"? And how do societies react when scientific discovery puts a damper on their narrative of exceptionalism? One of the most fundamental lessons from recent decades in human origins research is that there is no neat narrative where *Homo sapiens'* inherent superiority guaranteed our survival. Plenty of those pioneer populations that made their way into Eurasia went extinct. Perhaps rather than disparaging Neanderthals for not domesticating the reindeer they hunted, making metal bifaces, or building skyscrapers, we need to take a hard look at what we define as success in ourselves. Is the legacy we are on track for as a species one of massive, metastasized levels of material production and consumption, or of species-level collaboration built on empathy and altruism?

Being the last surviving hominin is not a victory, but rather a story of serendipity. Facing an uncertain future, it's time to accept that perhaps we didn't outlive the Neanderthals because of our capacity for cleverness or coercion, but thanks to our knack for conviviality, cooperation, and compassion. We can only hope that we share these attributes with lifeforms from worlds elsewhere, and that contact, whenever that happens, will be based on cordial curiosity, not fear and domination. ●

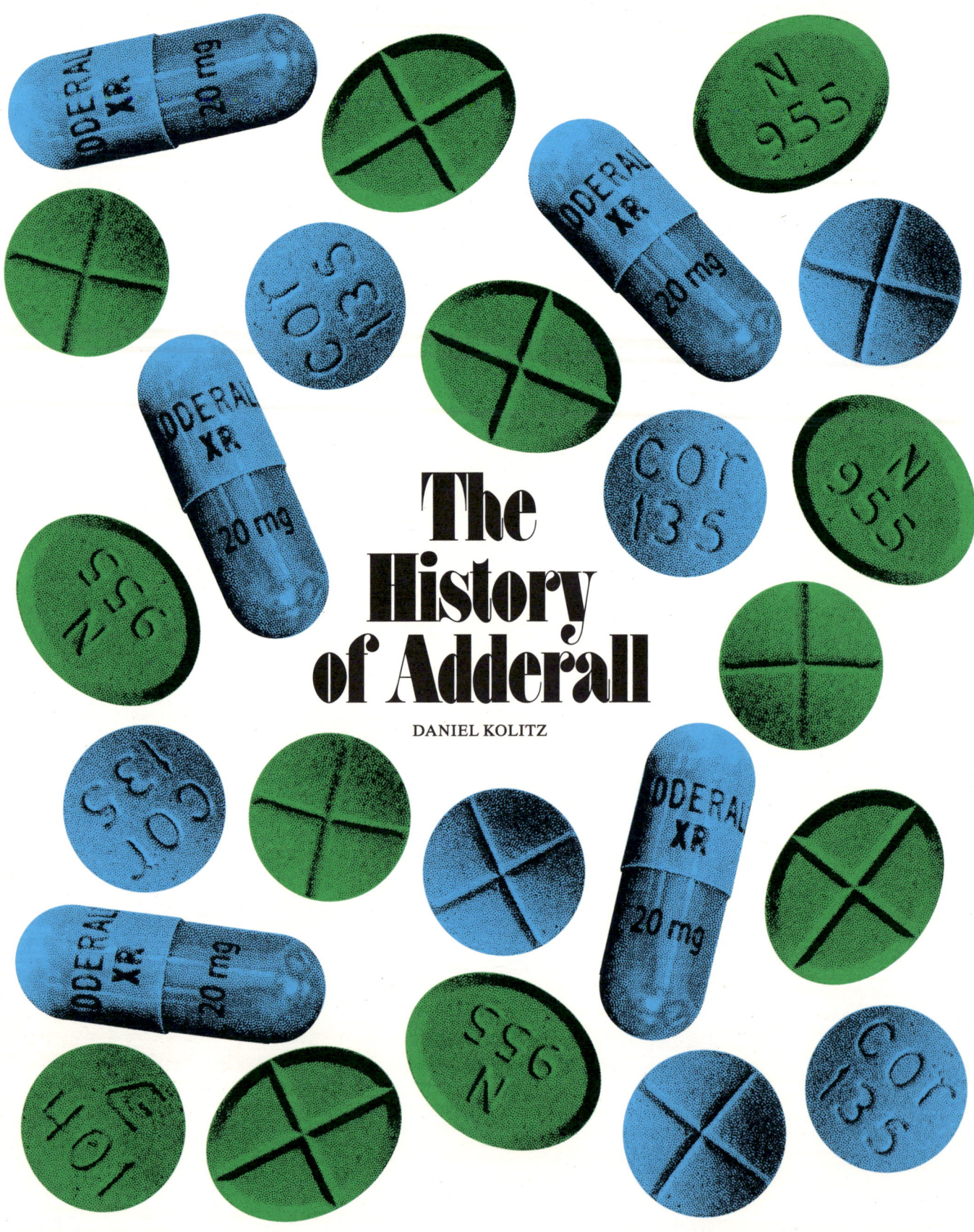

The History of Adderall

DANIEL KOLITZ

Since its approval by the FDA in 1996, the pharmaceutical known as Adderall has been relied on by millions to treat ADHD. This powerful amphetamine is also vastly over-prescribed, and its inducements to productivity, and ubiquity

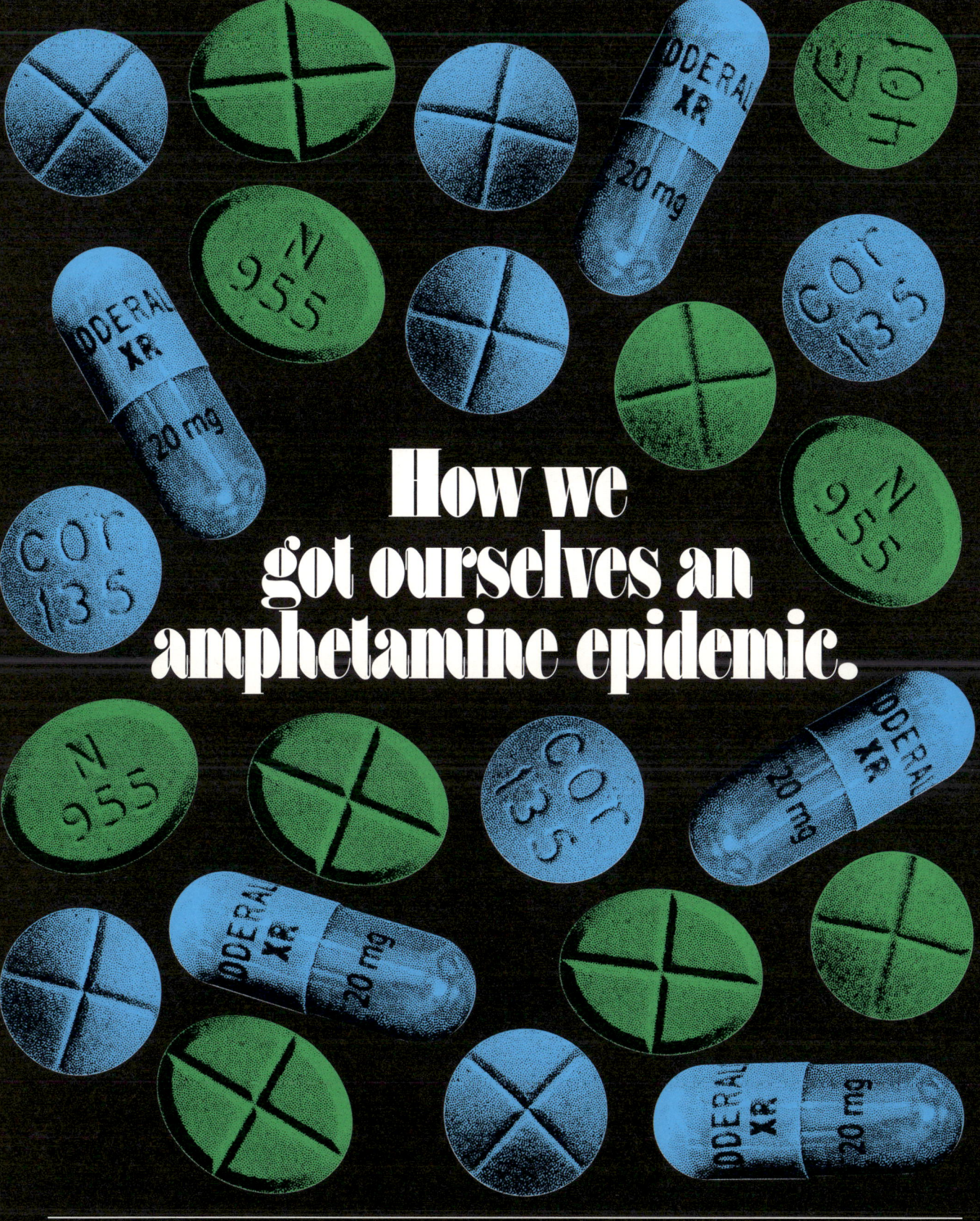

How we got ourselves an amphetamine epidemic.

among a certain sector of the professional class, have obscured many of its deleterious effects. The following is excerpted from "Club Med"—a collection of writings, published by *Broadcast*, exploring how Adderall has changed culture.

There was a time in my life, ten years ago now, when I did almost nothing but take Adderall and write about Adderall. These were complementary pursuits. The more Adderall I took, the more fervently I investigated Adderall—its culture, history, and collision with my life. While an alcoholic writing a history of alcohol might struggle to stay on the ball, in my case substance and subject were perfectly matched. Night after night I brought myself to the verge of a heart attack paging through abstruse FDA documents; mid-morning after mid-morning I went to bed assured of the rightness of my project. I was cornering the facts. I was blowing the doors wide open.

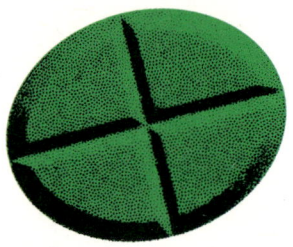

This project was personal to me. I was an early adopter, first prescribed Adderall in the third grade. At the time I felt that it was mostly poison. The come-up was great: there are layers to "Californication" you simply cannot hear unless you are nine years old, flying on pharmaceutical speed, and listening to it on your discman, ideally on the short-bus to Jewish day school. But the drug drained me of the one thing I cared about most: my prankish *joie de vivre*. No one likes a drugged class clown, except perhaps for that clown's teachers, school administrators, and parents. So I refused to take it. The jokes were more important. Later, a child psychiatrist would tell me that many people who hate Adderall as children wind up liking it in adulthood, after developing the kinds of worldly ambitions for which the drug was intended. He was right. At 23—a postgrad burning for self-advancement—I suddenly couldn't get enough of the stuff. You could almost say it was a problem.

Certainly, taking Adderall all night and drinking to fall asleep and then taking more Adderall to pretend to function at my largely fake job promoting various lifestyle products had its pleasures. But more and more often I just felt haunted and unwell. Even at the height of my mania I knew that this was no way to live. I also knew that I was far from the only person living this way. At hedge funds and media start-ups, coffee shops and WeWorks, the signs were plain to see: an entire generation grinding its teeth to dust, wrecking its posture, struggling to blink. How had we landed in this weird bind? Who exactly had allowed it to happen?

To help me answer those questions, I contacted preeminent speed historian Nicholas Rasmussen. His classic *On Speed: The Many Lives of Amphetamine* sheds valuable light on the period, in the 1950s and '60s, when amphetamine's ubiquity as an antidepressant and diet aid helped to usher in America's first speed epidemic. I was nervous to talk to him, partly because of his book's formidable intelligence and range, and partly because I'd taken a bunch of Adderall about 20 minutes before our scheduled Skype call. The call went poorly, as you can see from this excerpt:

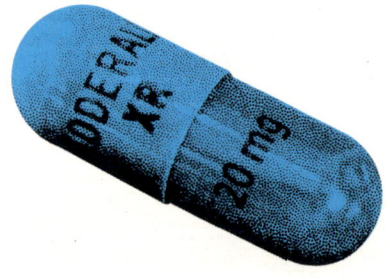

ME It seems, obviously, that most cases of amphetamine use don't ultimately end in psychosis or heroin addiction, although many do. And that's not to say that widespread amphetamine use isn't a social harm—but sometimes it seems like the consequences might be a bit more subtle or pernicious, especially when it's

being taken by, say, a working professional, who might actually be sticking to low doses, and not necessarily escalating and taking 10 then 15 then 20 milligrams…so I'm wondering what you think the downside of amphetamine is when it's used, maybe the word isn't responsibly, but when it's used in relatively low doses over an extended period of time.

RASMUSSEN I'm sorry—what?

ME I'm basically wondering, outside of the extreme cases of psychosis or heroin addiction, and escalating usage, what are the downsides of amphetamine use among the general population, if they're taking it in low doses over an extended period of time?

RASMUSSEN I believe that it's an addictive drug. So for those who don't get addicted—that is to say, those who only use it occasionally—there won't be any particularly great negative consequences. But since most people who are prescribed the drug seem to develop a dependence on it, at least adults, it strikes me as too dangerous to be used in this way.

ME So, what's the downside of being dependent on a drug like Adderall?

RASMUSSEN If you're dependent on amphetamine, you will develop amphetamine psychosis eventually.

ME Amphetamine psychosis just being defined as taking so much that you kind of go off the deep end, or—

RASMUSSEN It is a paranoid psychosis indistinguishable from schizophrenia.

Joke's on Rasmussen: I never went insane. Instead I developed a chronic tension headache that plagues me to this day, half-heartedly dabbled in the opioid epidemic (how else was I supposed to sleep?), and emerged, after ten months, with a strikingly unpublishable 12,000-word melange of memoir, broad-strokes cultural theorizing, and in-the-weeds regulatory analysis. In a way, the essay was a tortured account of the conditions—historical, political, neuro-pharmacological—that allowed for its creation. Reading it for the first time in almost a decade, I ache for the boy who wrote it. I also sympathize with his mission. Because he was onto something. The details of Adderall's FDA approval really are, I still think, a scandal.

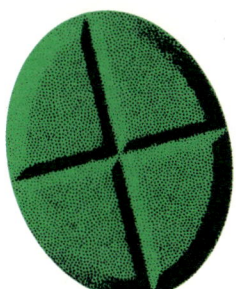

Adderall's precursor, Obetrol, was cooked up in a converted Queens garage sometime in the 1950s. Back then, pharmaceutical regulation was *laissez faire*. Becoming a large-scale amphetamine manufacturer was, more or less, a matter of buying the equipment and getting to work. You had to prove your drug was safe—a low threshold at the time— but no law stipulated that it had to work as advertised. Here's how William Goodrich, chief legal counsel at the FDA, described the situation:

The way drugs were investigated, a physician from the company would go out in the community with some samples and say to the doctor, "I've got this new drug for so-and-so. Here's some samples. Try it out and let us know how you like it." And they would get back a letter from him: "I tried it out on eight patients and they all got along fine." That's the kind of stuff that was coming in for science.

Goodrich was a virtuous public servant: on his retirement, he was hailed as the FDA's "most influential figure and guardian of its institutional memory." In an era when the agency was a public joke, lacking even basic enforcement powers, he made it his mission to harass profiteers—among them, notably, Rexar Pharmaceuticals, Obetrol's

manufacturers. In a complaint to the DOJ, he accused the company of selling high doses of Obetrol with no license, and promoting the drug with the "unsupportable and illogical" claim that it was safer and more effective than other amphetamines (a theme Adderall would pick up a few decades later).

The FDA finally banned Obetrol in 1973 as part of the agency's broader effort to curb the speed crisis, declaring the drug "ineffective" and "lacking in proof of safety." In response, Rexar reformulated Obetrol, swapping out the methamphetamine it contained for various amphetamine salts—the soon-to-be patented Adderall recipe—and resumed distribution, albeit in small enough quantities to evade regulatory notice. The drug's circulation in the 1970s and 1980s is hard to trace, although it had a way of popping up in huge quantities at random doctor's offices across the country. One news report in the 1980s highlighted the work of Ohio doctor Mattie Vaughn, whose practice was described in a court memo as "nothing more than a thinly disguised drug-trafficking operation," and who in 18 months dispensed nearly 140,000 of these new Obetrols to 4,500 patients.

By the early '90s, the market had largely dried up, Rexar's owner was dead, and his family had put the company up for sale. A Kentucky-based high school football coach-cum-pain pill entrepreneur named Roger Griggs examined its meager assets and seized on a surprising detail. Almost all of Obetrol's minimal sales—$40,000 a year—were attributable to a single child psychiatrist: a Dr. Robert Jones based in Provo, UT. Griggs made plans to visit him.

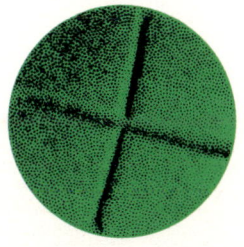

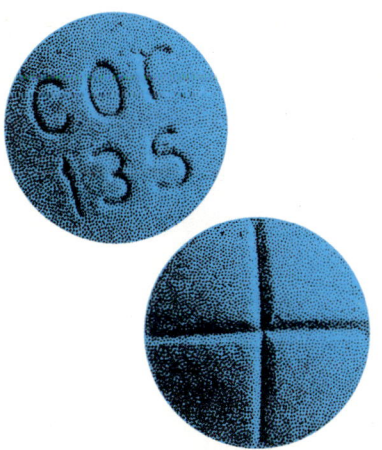

Over lunch in Provo, the enterprising doctor reported that he had been prescribing Obetrol to ADHD kids who couldn't tolerate Ritalin. The results, he claimed, had been spectacular. Griggs was sold. He bought the drug, changed its name (ADD for All: Adderall) and began marketing it at scale with his company Richwood. By the middle of 1994, just a few months into this campaign, 30,000 children were on it.

Unfortunately, Griggs had apparently skipped an important step. He soon received a call from the FDA, informing him that his flagship product had never received the agency's approval. It was "devastating," Griggs said, to learn their product was technically illegal. Nonetheless, Richwood promptly initiated a marketing campaign in a prominent child psychiatry journal, triggering a warning letter from the FDA restating that Adderall was an unapproved new drug, the continued marketing of which may result "in seizure and/or injunction."

There were good reasons to not approve the drug. Contra its early advertising, there was no evidence it was any more effective than Ritalin in treating ADHD. Its component parts had been generically available for decades. Asking for the "exclusive" right to manufacture Adderall was equivalent to asking for the "exclusive" right to manufacture a combination of peanut butter and jelly. The drug's chemical makeup and subjective effects were functionally indistinguishable from Dexedrine, which had lost its patent decades ago. Had Adderall been submitted as a completely new drug, rather than as a "supplement" to Obetrol's 1950s-era application, it would never have been approved—no company would even think to try in the first place.

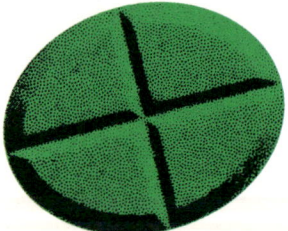

And yet—after months of mounting tensions—the FDA's tone abruptly brightened, and in a surprise about-face the drug was granted full approval and market exclusivity in January 1996, without a single substantial clinical trial submitted to demonstrate its effectiveness. Why? As *New York Times* journalist Alan Schwarz reports in his book *ADHD Nation*, a senator's son had been taking Adderall before the FDA intervened; the senator "pitched a fit" and the agency rushed to approve it. Who that senator was Schwarz doesn't say, but my research suggests it was Orrin Hatch, who took a curious interest in Richwood's plight around this time.

Sensing an opening, pharma giant Shire purchased Richwood for $185.7 million in 1997, and from there worked to advance a paradigm shift in psychiatry, one which would posit ADHD as an underdiagnosed scourge tanking the potential of not just millions of children but—once that market was saturated—millions of adults. In the meantime, Adderall has spawned multiple blockbuster sequels: Adderall XR, Vyvanse, and, most recently, Mydayis, a long-release version of Adderall which lasts, incomprehensibly, for 16 hours.

"We defined the attention deficit disorder market," Roger Griggs told me, ten years ago, as I struggled to extricate myself from both my dependency on Adderall and the mountain of Adderall-related documents,

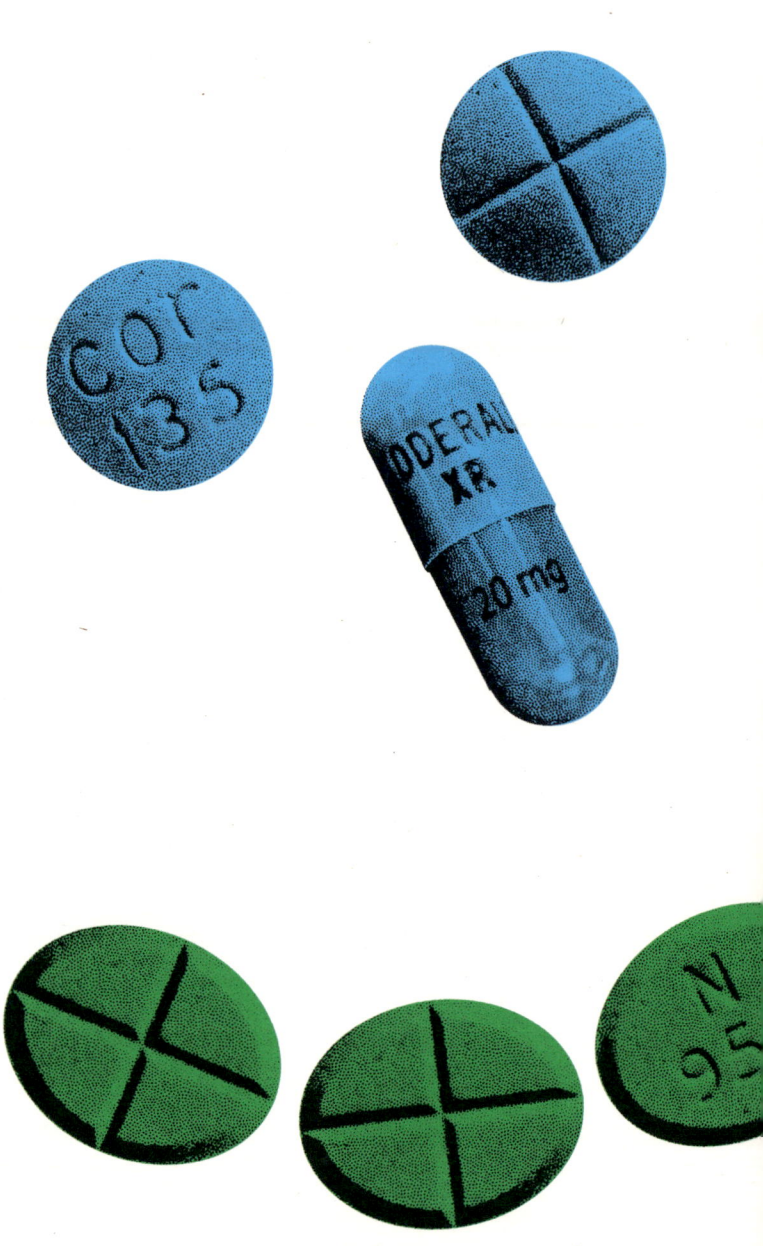

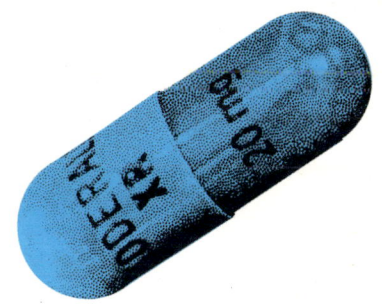

reports, statistics, and interview transcripts I had frantically compiled. "There was nobody promoting it. We were the first people to do that." By every conceivable metric, they succeeded. Per the most recent data, prescriptions of amphetamine—a chemical whose alleged therapeutic benefits were more or less entirely discredited at the start of the 1990s—have increased by roughly 2,500% since Adderall went on the market.

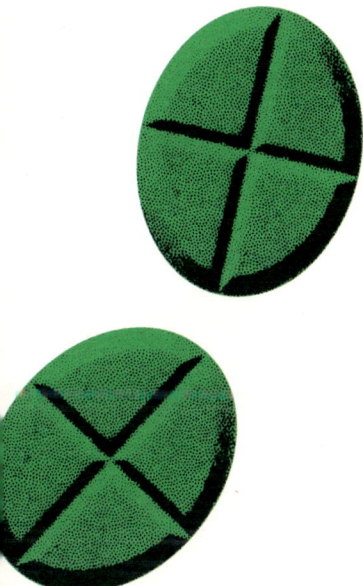

There is a robust, well-funded ADHD advocacy movement which insists that this is a good thing—that millions of people who would otherwise be struggling to read restaurant menus or walking blindly into traffic are now able, with Adderall, to live healthy, fulfilling, productive lives. And maybe they're right. User reports are so variable that generalizations are useless: few experiences are more epistemologically destabilizing than scrolling a heated Adderall-related Reddit thread.

I can say that, in my case, Adderall broadened my ambitions to the precise degree that it thwarted any possibility of achieving them. This had something to do with the quality of attention the drug generates, which is, above all, value-neutral. On Adderall, everything is interesting. This is one reason my Adderall essay was doomed from its inception. Whenever I seemed close to wrapping things up, another vital avenue of inquiry would materialize, until I

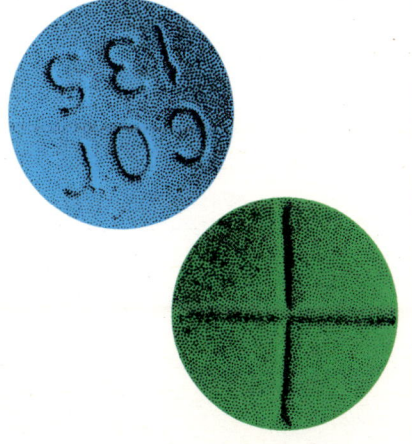

found myself—not entirely without reason—compiling material on minor figures surrounding the Kennedy assassination. Adderall, I eventually had to acknowledge, had destroyed my Adderall essay, and was on its way to destroying me.

The drug's influence has since escalated dramatically. As the recent shortages have demonstrated, there are hundreds of thousands of people who cannot function without it, owing perhaps to a deficiency of norepinephrine (the leading neurological theory for the basis of ADHD), or (as likely) to a form of iatrogenic chemical dependency, or both, or neither. Does it matter that the U.S. is the only country where this happens? That, however complex the rise in speed use, not to mention opioid use, might appear, these phenomena are effectively mono-causal?

The FDA approved OxyContin just one month before Adderall, and allowed Purdue Pharma to claim it actually *countered* addiction, thus creating that company's fatal marketing advantage. Likewise, Adderall was the first new amphetamine granted market exclusivity since the '60s speed epidemic—inexplicably, the agency granted Richwood a reason to get as many people as possible hooked on amphetamines. If Adderall's pushers have since evaded Sackler-style comeuppance, this can be largely attributed to the fact that it's basically impossible to fatally overdose on it. Entire desolated towns can testify to the destructiveness of OxyContin; all that Adderall leaves behind are the phantoms of lives that might have gone otherwise, my own included. ●

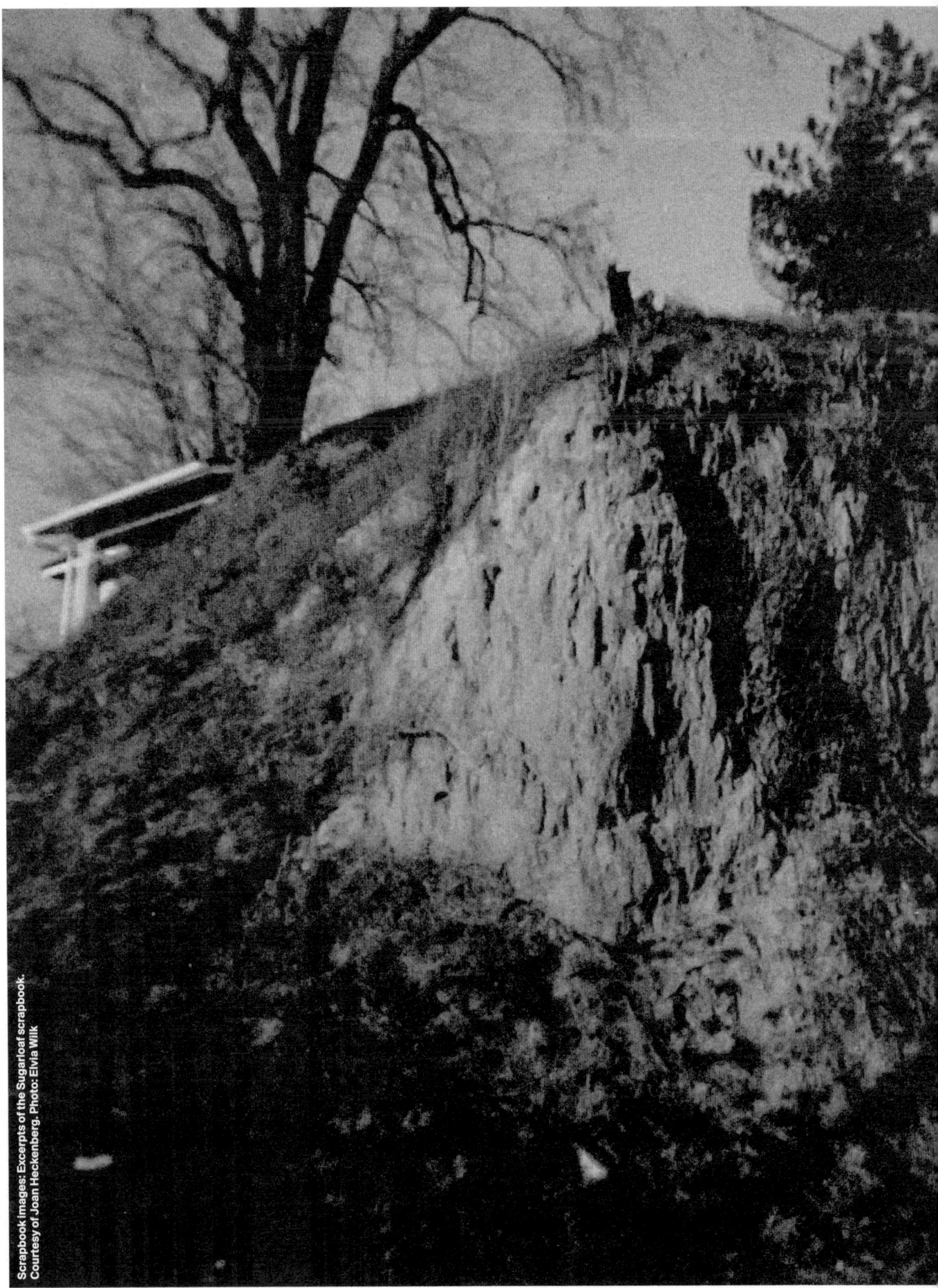

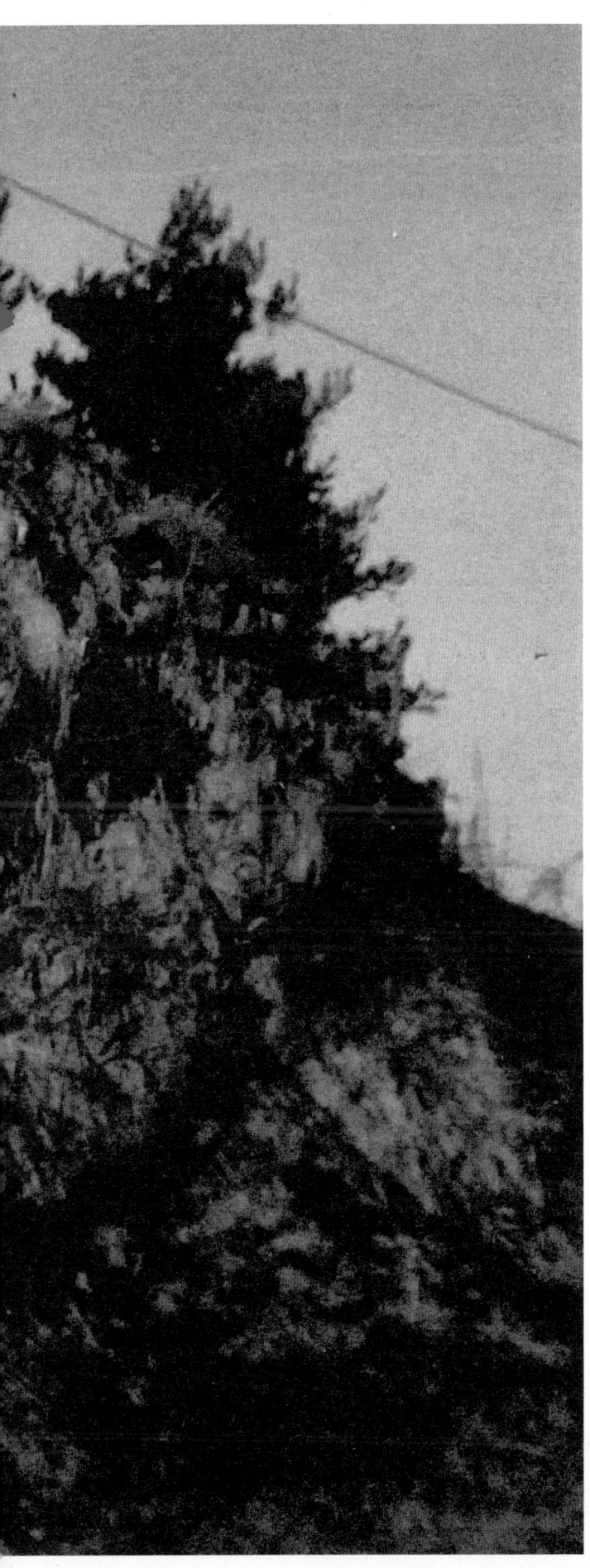

RETURNING SUGARLOAF

ELVIA WILK

Artists and curators in St. Louis are negotiating the transfer of a sacred Indigenous site. Can art world resources be leveraged for more than symbolic statements?

On a cold, sunny day in October 2022, two artists and a curator sat down in Joan Heckenberg's kitchen and offered to buy her house. Heckenberg's one-story, white clapboard home looks modest from the street, but its backyard has a wide view of the Mississippi River. The artists, Jackson Polys and Zack Khalil, had flown into St. Louis to visit her there twice before. The curator James McAnally dropped by so regularly that he sometimes helped out around the house. At this point, they had all gotten to know each other. The previous day, Polys had brought Heckenberg a jar of salmonberry jam from his family home in Alaska. Today, they brought her a cash offer of $160,000.

Heckenberg, a retired nurse in her mid-eighties, is used to receiving all kinds of visitors because she lives on what remains of a thousand-year-old sacred Indigenous site called Sugarloaf Mound. The Osage Nation has been working to regain ownership of the land for 15 years. In 2009, the tribe purchased a third of it, when the family who owned the bungalow atop the mound, just uphill from Joan, was ready to leave. Heckenberg has since been one of two remaining holdouts; the downhill property belongs to the Kappa Psi fraternity, associated with St. Louis's University of Health Sciences and Pharmacy. Over the years she has suggested that she plans to bequeath the land to the Osage in her will, but she has also mentioned a nephew who would like to inherit the house. At various points, the Osage have floated the idea of a sale, but Heckenberg hasn't wanted to leave her lifelong home.

In the course of their October visit, Polys, Khalil, and McAnally explained to Heckenberg that if she sold now, she wouldn't need to move out immediately—or at all—and laid out how this could be arranged. As core members of the group New Red Order, they have spent years documenting cases like this one, and interviewing people around the country about the various, and often ad hoc, ways that so-called Land Back transfers are made. One precedent for their current negotiation occurred in 2015, when a man named Bill Richardson decided to sell seven hundred acres of land on the California coast to the Pomo Tribe. The city of Oakland engineered a novel contract ensuring that he can continue living in his family home until his death, even though he technically doesn't own it anymore.

Land transfers to Indigenous groups, whether as gifts, sales, or bequeathments from individuals or institutions, are increasingly common in the United States, although they don't often make national headlines. In 2019, Eureka, a town in northern California, returned a two-hundred-acre island to the Wiyot Tribe, which had been seized from their ancestors during an 1860 massacre. In 2020, the nonprofit Conservation Fund bought twenty-eight thousand acres in Minnesota on behalf of the Ojibwe. As far as art institutions go, the best-known example is that of Yale Union, a contemporary art space in Portland, Oregon, whose board gave its building and land to the Native Arts and Cultures Foundation in 2020. Building on these successes, McAnally and New Red Order received a grant from the Andrew W. Mellon Foundation to support the return of Sugarloaf.

These efforts—often called *rematriation* to cast off the etymological baggage of patriarchal history—tend to involve years of negotiation. "The legal structure of America is set up to make it really hard to give land back to Indian people, and there's no clear way to do it," Zack Khalil explained to me. The movement for territorial sovereignty has loosely coalesced under the term Land Back, but it remains decentralized. Collecting and comparing stories makes the process easier each time, sparing people on both sides of land transfers from "reinventing the wheel every time." And yet the circumstances here were somewhat unprecedented: artists were helping broker a significant rematriation project, in the context of a contemporary art triennial, called Counterpublic, backed by a private foundation.

And it appeared to be working. After a decade of wavering, Heckenberg seemed to be making peace with the idea of a sale, and even with the prospect of moving out, too. Everyone in the room was emotional. "In this surreal conversation," McAnally said, "she was very much talking in the past tense, as if she had decided to leave. There was this whole performance of *this is the ending*." But then Heckenberg switched gears. She said she didn't want to leave until her elderly dog died. She added that she might want to bury the dog on the mound. "And then with this little wink," McAnally recounted, she said, "I don't think the Indians would be too happy about that." The comment may have been off-kilter, but it was in keeping with her sense of humor. They left cautiously hopeful. They would give her some space and then come back.

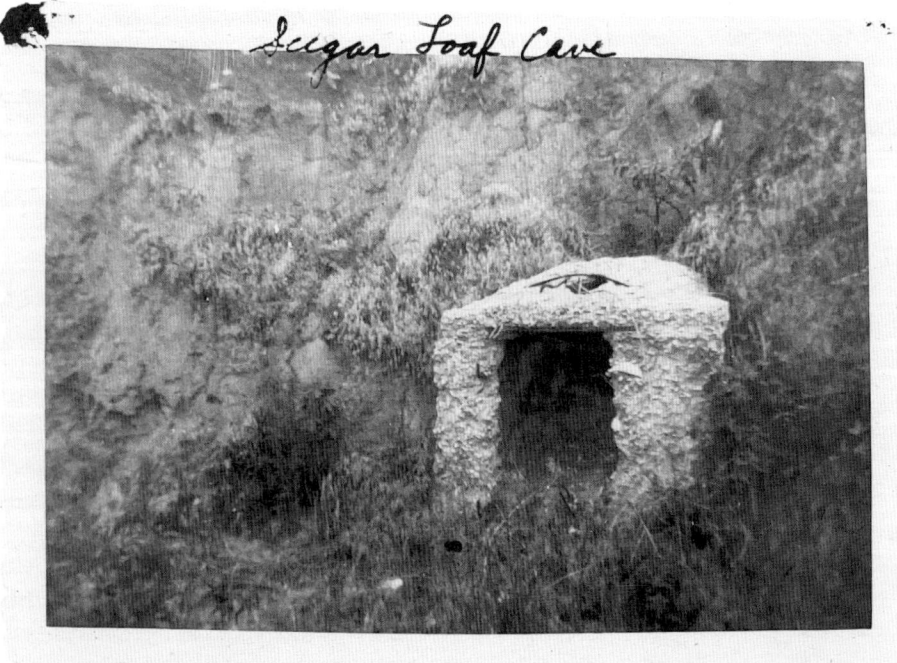

Sugar Loaf Cave

etween 800 and 1600 AD, the area around what's now St. Louis was home to the largest precolonial settlement north of MesoAmerica. Here, at the intersection of the Mississippi, Missouri, and Illinois rivers, the Mississippian civilization presided over a thriving hub of trade and agriculture. These groups constructed dozens of platform mounds, impressive earthworks that were used for various purposes: ceremonies, astronomical observations, residences, burials, and sending signals, such as fire signals, over long distances.

Several tribes, related by the Dhegiha Siouan language group, were living among the mounds when, in the 1670s, French explorers started to float down the rivers in canoes. We know who was living there in part because the French wrote the tribes' names on maps and documents: the Osage, Ponca, Quapaw, Omaha, and Kaw. Soon, French, Spanish, and British settlers caught on; this was a great place to live. In 1682, a fur trader decided that the area would now be part of France. He named it *La Louisiane*.

In their foundational text "Decolonization is not a metaphor," which has become a touchstone for Land Back movements in the United States and beyond, Eve Tuck and K. Wayne Yang describe the particular form of violence that followed first contact in the Americas. "Settler colonialism is different from other forms of colonialism," they write, "in that settlers come with the intention of making a new home on the land… Land is what is most valuable, contested, required. This is both because the settlers make Indigenous land their new home and source of capital, and also because the disruption of Indigenous relationships to land represents a profound epistemic, ontological, cosmological violence."

In 1803, the newly formed United States bought a third of North America for $15 million in the Louisiana Purchase, and St. Louis became a crucial waypoint—dubbed "Mound City" and "the Gateway to the West." For the next century the river basin was the heart of colonial expansion. It became a railway transit point, an expanding urban center, and home to the armories and weapons caches necessary to kill or displace all the people living in the "empty" territory west of the river. Over the course of several waves of slaughter and displacement, settlers began to raze the mounds, flattening and parceling the landscape. In 1825 the Missouri Osage made a treaty with the United States government, forcing them to move to Kansas. In 1871, they were again relocated, this time to Oklahoma. From 1845 to 1909, native people were legally banned from living within the St. Louis region. Today there are no state or federally recognized tribes in Missouri.

Sugarloaf—named so by the French because it reminded them of the cones they used to transport sugar—is the last remaining mound on the Missouri side of

the river. On the Illinois side, the Cahokia Mound complex stands relatively well preserved. Furred with mown grass, it is open to visitors, who can see what the mound may have looked like a thousand years ago through an augmented reality app. Cahokia was declared a UNESCO heritage site in 1982.

Sugarloaf has only survived until today thanks to a string of historical accidents. In the nineteenth century it was damaged by a quarry near its south side, and a chunk of the mound was carved away for a dirt road. In the 1920s, families started settling on the still-stable parts of the mound, building three houses from its highest point to its lowest tier. Heckenberg's grandparents built a house on the middle plateau and she moved in with her parents when she was about five years old. In the 1960s, the I-55 interstate was planned, and would have cut directly across Sugarloaf—but then a land survey showed that mining had made the ground unstable, and the highway was nudged three hundred feet away, incidentally sparing the houses. With brief gaps, Heckenberg has lived in the middle house for 80 years.

ecruiting people like Heckenberg to just "give it back" is central to New Red Order's art practice. Founding members Jackson Polys, Zack Khalil, and Zack's older brother Adam Khalil sometimes describe themselves as a "public secret society." Their name is a sardonic reference to the Improved Order of Red Men, a still-existent fraternity first formed in 1813 ("whites only" until 1974) that claims lineage to the Sons of Liberty, who in turn claimed responsibility for the Boston Tea Party. One of the artists' longest-running projects is a recruitment campaign for "non-native informants"—corporate-style videos and a website advertising a working hotline that allows non-native callers to dial in, inform on settler culture, and join NRO. Their primary recruit, who has become something of a mascot, is veteran actor Jim Fletcher, a "reformed Native American impersonator," who appeared in a 2015 Wooster Group performance in Indian costume. He now features in NRO performances and videos delivering (deadpan, funny) apologies and invites others like him to "change your life by learning to recognize—and report on—the efforts of non-Indigenous people everywhere to claim indigeneity."

The materials of the ongoing "Give It Back" campaign lampoon corporate aesthetics and feel-good lingo. NRO are emphatic about looking for "accomplices, not allies." This vocabulary is drawn from a 2014 pamphlet by a collective called Indigenous Action, and proposes that allyship of the lip-service "solidarity" type can exploit cultural signifiers to distract from the real terrain of struggle, or absolve people in power from making real material change. Accomplice-ship, meanwhile, is focused on direct action.

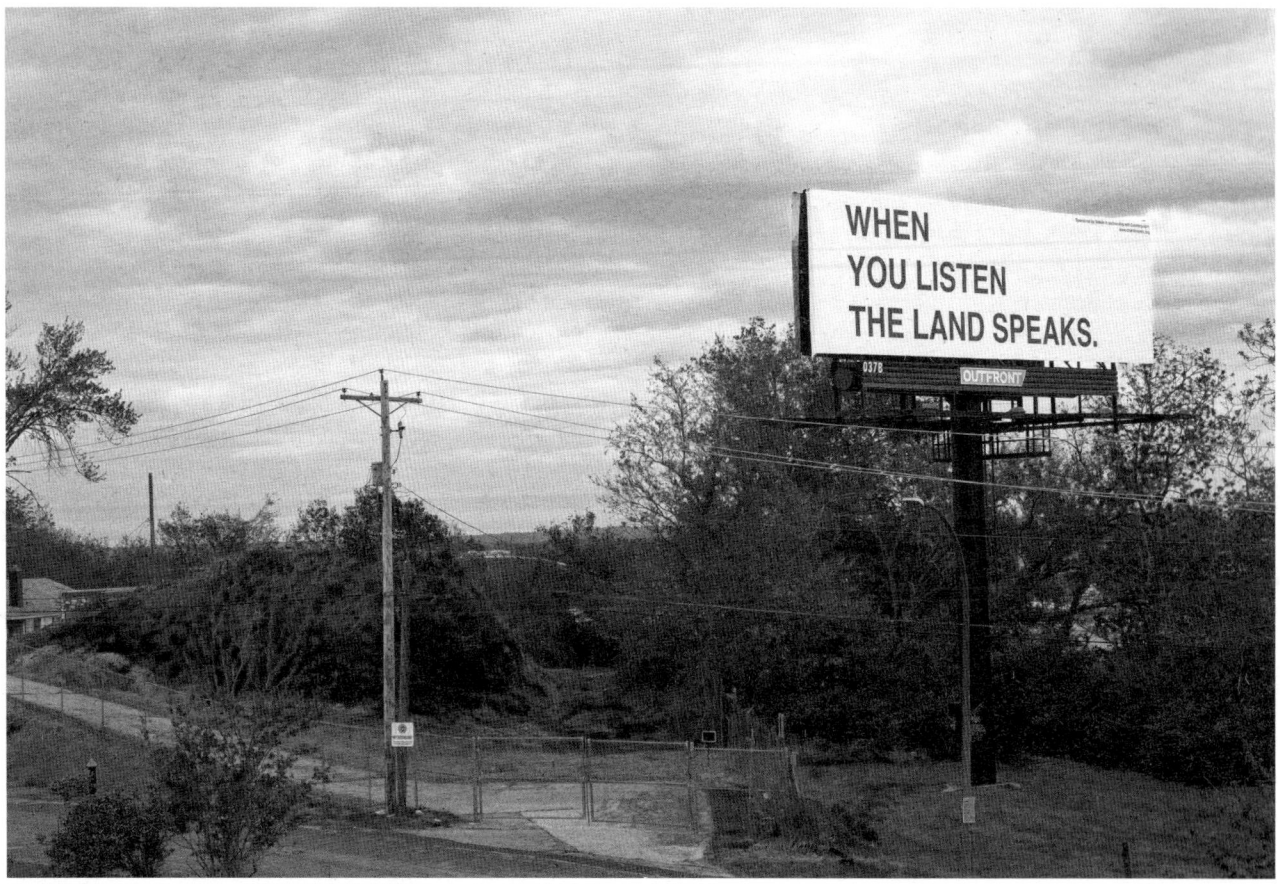

Anna Tsouhlarakis, *The Native Guide Project: STL* (2023), mixed media. Photo: Jon Gitchoff

I went to college with the Khalils and have followed their work since. In 2016, the brothers, who are Ojibwe from Michigan's upper peninsula, made the film *INAATE/SE/*, a mix of documentary and fiction that imports an Ojibwe first-contact prophecy into the present day. Around that time, they met Polys, a Tlingit artist particularly well known for his carved sculptures, and the three of them started talking about how they felt that they had been conscripted into a "native informants' role." That is, they found themselves tasked with interpreting and sharing native cultures with largely white art audiences—extracting from their own lives and histories for show. In 2018 the three "inducted" themselves into New Red Order (although their "real induction," Adam told me, happened in 1492). Their first piece together was a video about the "Kennewick Man," a nine-thousand-year-old Indigenous human skeleton that the Colville Tribes spent 20 years petitioning to get back from the National Parks Service, the Army Corps of Engineers, and the Smithsonian Institute.

NRO's increasing notoriety is due to several expansive projects in sites far outside the museum, with collaborators far beyond the art world. For their largest work to date, a 2023 Creative Time commission, they flipped the classic colonial World's Fair model and built The World's UnFair in an empty lot in Long Island City. In the center of the weedy lot, a goofy, disconcerting animatronic tree spoke with a goofy, disconcerting animatronic beaver about the history of private property (*Dexter and Sinister*, 2023); a five-channel video installation broadcast Jim Fletcher's exhortations (*Give It Back*, 2023); and a stage hosted revolving conversations and performances. NRO's projects often seem deceptively ad hoc or casual, when they in fact result from intensive research, organizing, and recruiting. Asked about this playful approach, the Khalils refer to the "trickster element" of the Ojibwe cultural tradition.

When Adam Khalil—holding a red Solo cup and wearing his signature NRO-branded wide-brim cap—first mentioned the Sugarloaf project to me in 2022, I was not surprised by the scale of ambition. NRO has long insisted that art-world resources can be leveraged for material change rather than symbolic statements. At one point, they approached curators at the Whitney and asked whether the museum would consider handing over its Met Breuer building to an Indigenous collective. (In 2023 the Whitney sold it to Sotheby's instead.) But I was surprised that Mellon had offered money for a Land Back cause—major funders can be skittish about such sensitive initiatives—and impressed that NRO had made such a leap across the provocation-action divide. It's one thing to run a tongue-in-cheek "Give It Back" campaign. It's another to secure hundreds of thousands of dollars to get it back.

James McAnally, the initiator of the renewed Sugarloaf effort, never intended to start a big triennial. "The formation of institutions is what's interesting to me," he told me in a conversation at my kitchen table. It was only after trying lots of other models aimed at bringing art people and everyone else together that he ended up founding Counterpublic, the triennial gathering he's coordinated in St. Louis since circa 2018. His reluctance was overcome by his sense of what such a structure—compromised, flawed—could achieve.

Everyone hates, or claims to hate, the -ennial model. If you read art magazines, you know that a large percentage of reviews of these periodic events begin with a cursory paragraph about how fucked up the premise is (I have written such paragraphs myself). The problem, writers point out, is that biennials and triennials tend to be temporary, extractive, apolitical interventions into local landscapes with little regard for what's already happening there. Curators are usually imported from elsewhere and assembled based on international pedigree rather than local knowledge; the main audience is the international press. At their worst, these events function as shopping malls masquerading as critical interventions, like tiny Olympic games that the general public doesn't care about.

Counterpublic, on the other hand, grew out of a highly localized DIY scene and was aimed at long-term engagement. McAnally, an artist, musician, and writer from Mississippi who moved to St. Louis for college and afterward ran a popular art space called The Luminary, launched the progenitor of Counterpublic in 2018—an exhibition whose venues included a barbershop, a Buddhist temple, and a taqueria. The project grew naturally from the following year's 24-site show, which the curators called "a triennial scaled to a neighborhood," to 2023's three-month exhibition sprawling across 13 neighborhoods.

More than half of the Counterpublic budget was given to people, groups, and institutions in St. Louis. A large portion went to the Griot Museum of Black History, some to the more recent George B. Vashon Museum of African American History. One artist, Maya Stovall, decided to give part of her project budget away as an act of reparation, in an undocumented piece called *Theorem, no. 3*, which preserves "the trust and anonymity of participants in no-strings transmissions of the official currency of the United States of America." Counterpublic commissioned 37 artworks, some of them permanent. One was a monumental public sculpture by Damon Davis that memorialized Mill Creek Valley, a central neighborhood of twenty thousand residents, mostly Black, who were displaced when the entire area was razed in 1959.

For years, McAnally had wanted to do a project involving Sugarloaf, and 2023's Counterpublic helped him amass the necessary infrastructure. Counterpublic brought in Risa Puleo, a contemporary art curator and art historian of Mississippian cultures, who laid the ethical groundwork for approaching the mound. The two wrote a letter to the Osage Nation Historic Preservation Office asking to use the mound as a central exhibition site. They pledged a "high profile opportunity for community engagement"; to push for "civic action"; and to "materially support" the land transfer. Once the Osage's cultural advisers replied with a go-ahead, McAnally enlisted NRO as participating artists and curators. Zack Khalil and Jackson Polys began making trips to St. Louis—Adam Khalil was abroad—researching the area, communicating with Osage preservation officers, driving to Heckenberg's house and ringing the bell at her gate. "Initially," Zack said, describing their attitude going in, "there was an understanding that Joan wanted to give the land back." But "it became apparent through speaking with her that it wasn't clear a transfer was happening, so that changed the dynamic of the conversation."

"It came down to money at some point," said McAnally of the dialogue with Heckenberg. When the Mellon Foundation invited him to submit a project grant for Counterpublic, rematriating Sugarloaf was included as a line item in the budget. Since the poet Elizabeth Alexander became the Mellon's president, in 2018, the grant-maker has placed a new emphasis on monuments and memory. Counterpublic received a $2 million grant to support the civic exhibition, with $350,000 earmarked solely for the mound. The funds were contextualized within the framework of monument protection—a framework and tool most often used for heritage preservation and nature conservation, but which has come to be seen, by some, as a productively roundabout way to advance Land Back efforts like Sugarloaf.

At that point McAnally, Polys, and Khalil were ready to visit Heckenberg with their $160,000 offer, based on a rough appraisal of her house. In a letter McAnally brought her that day, he clarified: "What this means for

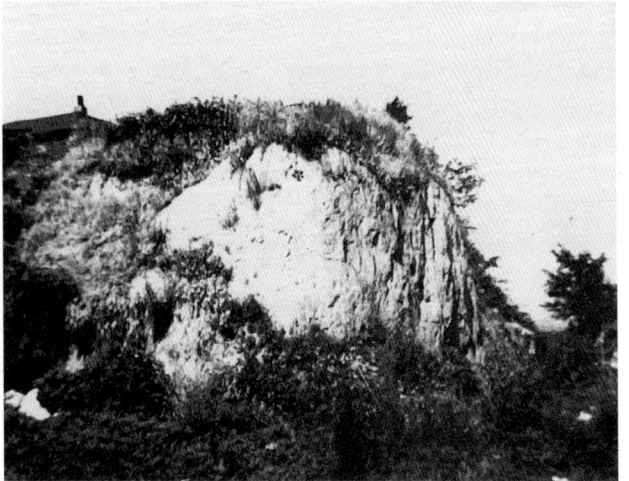

you is that we would be able to compensate you for the transfer of your home… Like you, we agree that this is the right thing to do for all involved, and have an urgency to complete this process."

At no point, Khalil and Polys both emphasize, did anyone push Heckenberg to leave. Polys explained that if there was any element of coercion, "it could have been us who are doing the dispossessing. We don't want that. We want to allow for a different kind of model and relationship building." This was to say that Heckenberg, too, could at any moment accept the invitation to become an accomplice.

St. Louisans, if they know about the mound at all, are largely unaware that it has not been fully rematriated. The city flaunts Sugarloaf as a success story. At the Museum of the Gateway Arch, a wall text introducing the main exhibition proudly announces that "the Osage nation owns and preserves Sugarloaf." The Counterpublic cohort has worked to revise this misconception. For the 2023 exhibition, they replaced the advertisements on either side of the billboard adjacent to Sugarloaf, which usually advertise a weed dispensary and a personal injury lawyer, with two artworks by NRO and the Navajo-Creek artist Anna Tsouhlarakis. The NRO sign, which faces Heckenburg's house, announces in colorful, cartoonish lettering: "THIS BILLBOARD IS ON SACRED LAND," "GIVE IT BACK," and "NEVER SETTLE," over a painting of an idyllic landscape with a mound-like earthwork, and a house that looks a little like Heckenberg's. Tsouhlarakis's black-and-white sign reads: "WHEN / YOU LISTEN / THE LAND SPEAKS." At the base of the mound, near where the quarry used to be, Osage artist Anita Fields installed forty wooden platforms resembling those used at Osage events today to create a gathering place, resulting from curator Puleo's invitation to members of the Osage to engage with the site.

McAnally has been urging city officials for years to consider supporting the transfer in a public statement. When I asked him what a city could do besides proclaiming solidarity, he pointed out the obvious: a municipality can use eminent domain to seize any property it deems culturally or economically significant. I had forgotten that transfer could be so straightforward—and that the personal (what Heckenberg does with her house) is always structural (she's there because laws defend her right to be there).

But even when land is given rather than sold, it often has to be valued first. Priceless heritage must be priced so that it can change hands and be converted to something like pricelessness. The language of property has a way of obfuscating what's really at stake: sovereignty—the kind of sovereignty that operates beyond property deeds registered with current governments.

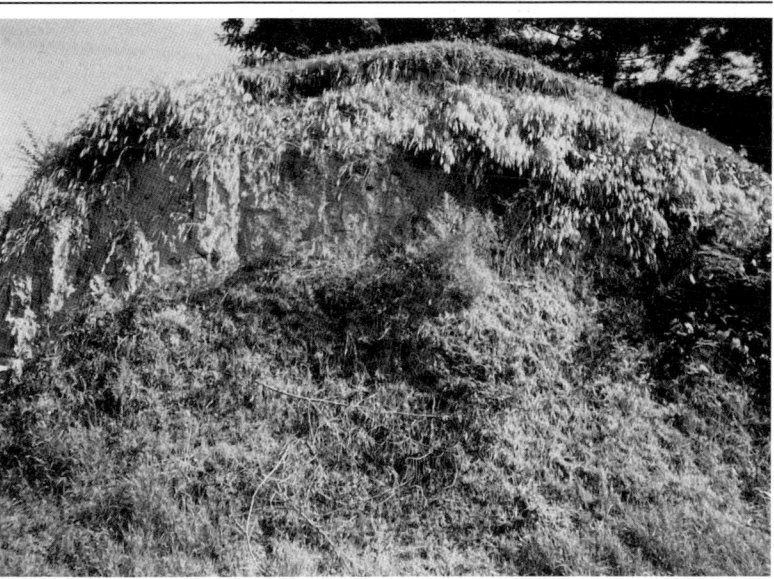

Driving south on I-55 from central St. Louis, you probably wouldn't notice the mound. The 40-foot-high bluff is barely visible from the raised highway running alongside it. I wouldn't have known where to turn if I hadn't seen the enormous billboard rising up across the lanes. I traveled to Counterpublic during its July 2023 closing weekend, and after reading and hearing so much about the mound, I was surprised by how small it looked—just a grassy hill, really. A roughly paved road called Ohio Avenue leads from the highway turnoff to the two remaining houses. A chain-link fence designates the rematriated area at the top. Just below the ridge is Heckenberg's unassuming house, her front porch decked out with folding chairs.

By then, I was frustrated by how outsized a part of this story Heckenberg had become. I had become slightly obsessed with her, and was angry at myself for being obsessed with her. I wanted to write about the role of artists and cultural institutions in making material change, and the hard work of decolonizing a continent piece by piece—and here I was, pursuing an octogenarian who I gathered might have a case of main character syndrome. At the same time, she was a person, not a character, and I didn't want her to flatten into one in my mind.

I didn't have an appointment, but I had heard that she was usually home and liked visitors. I tentatively rang the bell tied onto her chained gate until she emerged from the house, wearing loose pajamas with paint smears. She didn't look surprised to see me. "I'm painting some of the basement walls," she said, gesturing to her paint-speckled clothes. She asked what I wanted to discuss with her, although it was obvious. I wanted what everyone wanted: to ask about her house. A breeze lifted a fluff of white hair from her forehead, and a little dog ran out from behind her, also fluffed with white hair. "I can talk to you for a little bit," she said casually. I stayed for two hours.

Heckenberg was kind. She invited me to sit on her front porch and offered a soda. She is thin but not frail, although she complained that her whole body is "eroding, just like" the cliff behind her house. Trucks whizzed by constantly on the highway just across the road, shaking the pavement beneath my feet. Her yard was mowed and a little Virgin Mary statuette was planted near the front fence. She told me someone else now tends to her lawn, because the last time she did it herself it looked like it had a "Mohawk haircut."

Her memory is extremely sharp. She can describe what's happened on the mound in detail—always mentioning people by their first names as if I knew who they were, in the same confusing way that my own grandmother used to. She told me that as a child, she'd "played cowboys" up on the hill where the house of her neighbors, the Strosnider family, used to be. Back then, she and her parents thought they were living on a burial mound, but now she knows it's probably a signal mound. In 2015, Dr. Andrea Hunter, an archaeologist who directs the Osage Nation Historic Preservation Office—who also led the task force to rematriate the first third of the mound back in 2010—brought a team to investigate the soil. They did remote sensing using a radar device that Heckenberg said "looks like a golf cart," and found no evidence of human remains or artifacts.

She brought out her family scrapbook to show me her mother's collection of photographs and newspaper clippings, which she's added to. The pieces are lovingly assembled and annotated with descriptions and dates—"my front yard" or "the old quarry." There's a letter from the Missouri Department of Natural Resources thanking Heckenberg for her "efforts toward the preservation of this significant historic site," upon Sugarloaf's designation entry into the National Registry of Historic Places in 1984. There's a page with cut-out drawings of the family of Sauk chief Keokuk, who was born around 1780 about 150 miles north along the Illinois River. There are pictures of a nearby workhouse before it was torn down in the '50s. Heckenberg explained that the foundations of her house were built by prisoners who lived in the workhouse.

About half of our conversation was preoccupied, not with the Osage or Counterpublic, but with the "hobos and transients," "hookers and homeless people," who often travel over or stay on the mound. The mound is cut off from the city by the highway and its backyard overlooks train tracks by the river; it's hardly well protected. For years, a mentally ill man squatted in the frat house down the hill with two pitbulls; he stole her mail, went through her trash, and threatened her. She joked that her dog, Molly, who is 17 years old and blind, isn't much of a guard dog anymore. The way Heckenberg tells it, living alone and vulnerable, armed with prayers and a gun, is stressful and frightening. Finally, she said with relief, the frat has lent the ramshackle home to a Palestinian man, who's agreed to fix it up and keep squatters away in exchange for free rent. (I didn't knock on his door due to the large sign saying "TRESPASSERS WILL BE SHOT.")

Zack Khalil told me that Heckenberg's preoccupation with her safety is one reason why they thought she might want to move. "She's pretty isolated up there," he said. "It's like, okay, is there some way that we can help this person? It seems like kind of a win-win situation." At one point, Counterpublic helped Joan tour condominiums that she could afford with the purchase money, but she didn't like them. "I don't want to buy a house," she said to me. "I'm too old. See, when you're young, that's fun. You pick out your drapes, what color you want the walls painted. When you're 85 years old… I don't want to redecorate."

I asked her in just about every way I could think of whether she believed that native people were entitled to the land. She'd say things like: "Oh, absolutely. Yeah. Oh, sure." She worked in conditionals—what *would* or *should* happen.

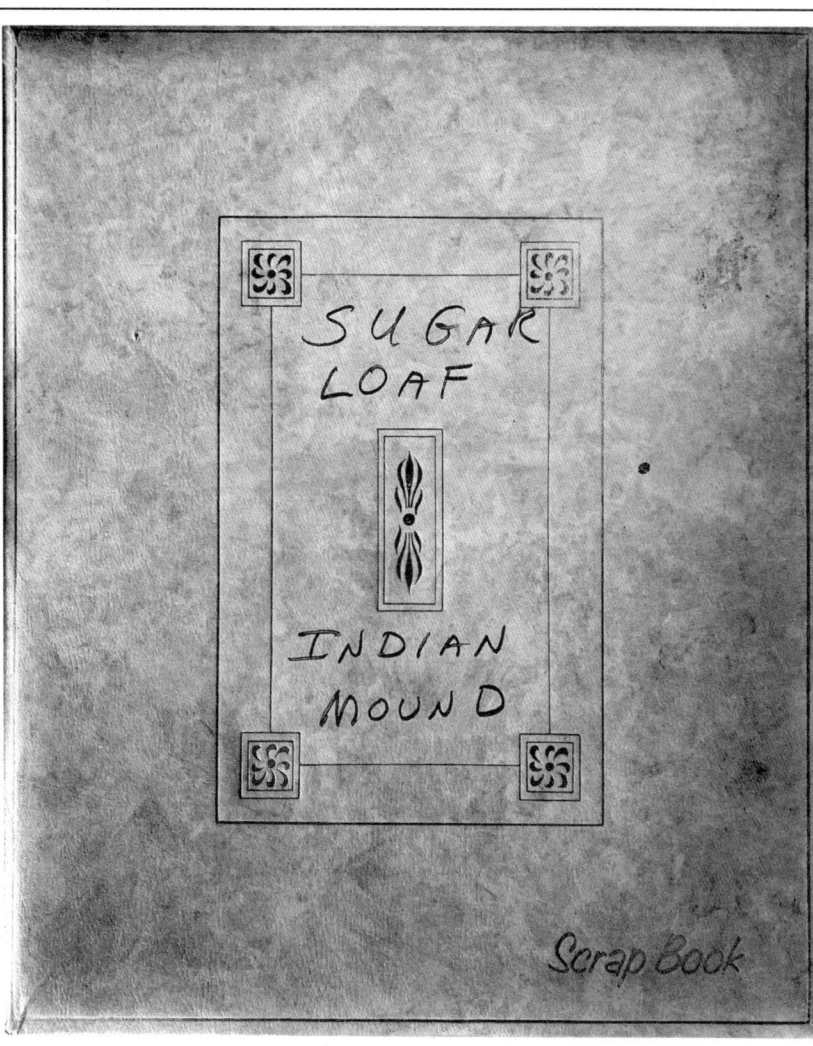

Sometimes she pressed fast forward to what *will* happen without explaining how it will come to be, or talked in the past tense as if a handover had already happened. I remembered something McAnally had said about her way of speaking: "At first it was kind of like a quirk of her personality… and then it became clear that it was a strategy. She's very savvy."

"I am not moving until this dog's gone," she said, her voice rising with frustration, in protest against someone not there. "I am not standing in January in the freezing rain with an umbrella at three in the morning waiting for a dog to go to the bathroom. I see people do that. Yeah, I have seen people in terrible weather, older people standing with the umbrella for the dog to move its bowels, which sometimes takes a little while. I'm not doing that at my age, I let her out three in the morning. She goes out a lot at night. She's old. When you get old, you gotta go to the bathroom a lot. She goes out, she pees, she comes back in, we go back to bed. I don't have to put shoes on, and a coat and gloves. I'm not doing that at my age. When she's gone, we'll take it from there."

Before leaving Heckenberg's house, I pointed up at the billboard designed by NRO, which she can read from her yard. "THIS BILLBOARD IS ON SACRED LAND. GIVE IT BACK." I asked her what it was like to see the sign all the time. She shrugged as if she didn't know what I was asking, and said it was awfully bright at night.

At the end of 2022, with Heckenberg continuing to prevaricate, Counterpublic reconsidered the Kappa Psi property, a brick duplex and garage built in 1968 and acquired by the fraternity in the '90s. Frat bros used to live and party here, but Kappa Psi—"the oldest and largest pharmaceutical fraternity," says their website—has left the home vacant for many years, seemingly disinterested in its upkeep. Neither current members nor alumni have been remotely interested in dialogue with the Osage, much less contemporary artists. Early on in the curatorial process, McAnally said, "We approached them first as an arts organization saying, we are looking at this property for an art project. Full stop. That got us a response at first, but it didn't lead anywhere."

With funds from the Mellon Foundation in hand, they could make a formal offer on the frat property as well. The house was officially appraised at $175,000, and Counterpublic tried offering $160,000, then raised it twice, ending up at $220,000. "Their only response to date is that they want us to buy them condos closer to the college campus," McAnally told me, "and that would cost like $400,000." When I spoke with Dr. Hunter, she said, "To be honest with you, I just don't get it. They've been flat out rude to us." Heckenberg told me that she is also baffled about the frat's refusal to consider the sale: "It used to be such a nice house. They could have sold it to ordinary people."

The frat may be stonewalling because they know Indigenous land is worth more than it's *worth*. In 2021, the beer-making Busch family (net worth $17.6 billion) decided to sell a piece of their property about sixty miles west of St. Louis. The property includes an underground rock art site known as Picture Cave, with about three hundred magnificently preserved Mississippian paintings from a thousand years ago. Finally, here was an opportunity to buy it back; the Osage negotiated with the Busch family for months. "The more we talked about it, the price kept going higher," Dr. Hunter told me. "They knew the importance of this cave and the artwork that's in it. We kept having to go back to our Osage Nation congress asking for money, and then they got impatient and said they were tired of waiting on us to get the funds and they were going to auction." The site eventually sold at auction to Morning Star and His Friends LLC, a Nashville-based company with two unnamed members and a law firm as its address. Nobody has been able to figure out who owns the company—or the cave it's acquired.

When I first called Dr. Hunter in September 2023, she was wary. "Why are you writing this?" she asked. She asked me to tell her about my experience with tribes and to explain what I knew about the Osage. It's clear that rematriation is extremely sensitive, but I wasn't sure exactly what she was so guarded about—until she told me what types of cold calls she typically fields. There are "people calling us and wanting me to come look at their property because they think they've got a pyramid buried in their property that's glowing," and "people wanting to do a movie about aliens and one of them is going to be an Osage that crashes on our reservation."

At the outset, I gathered that she had regarded Counterpublic's first outreach with similar wariness. "I thought this is crazy, what do artists have to do in our business? I really wasn't thrilled about it in the beginning. People have odd interpretations of tribal people and there is such a huge lack of information and misinformation about tribes." But after several conversations, Dr. Hunter's team invited Counterpublic to submit a letter outlining their proposed work with Sugarloaf, and in turn, tribal leaders unanimously agreed. Since then, Counterpublic has submitted all artwork and writing regarding the mound to the Osage for review before acting.

Most Osage revenue comes from the casinos on their reservation, and requesting funds for rematriation of any kind goes through a formal congressional procedure that can take longer than, say, the Busch family is willing to wait. Allowing the triennial to act as a go-between seemed like it could speed things up. "I thought, all right, fine, I'll respond to them. So I just said, if you think you can help in getting Joan's property and getting this fraternity's property, have at it. We've been trying to do it for years with very little success. I figured it wouldn't hurt to have them give it a try."

To add a layer of complication, the question of whom to give land back to is not always clear-cut. Waves of displacement over centuries mean that tribes have now lived for generations on land that wasn't "originally" theirs. Dr. Hunter and others have gathered evidence from oral histories and artifacts, to demonstrate the long presence of the Osage in the area. And still this makes no difference when private property is at auction: a seller can choose any buyer they want, as with Picture Cave— yet "another painful experience."

She emphasized the uniqueness of Sugarloaf, explaining that the Osage mainly buy back land for their own reservation in Oklahoma rather than working out-of-state. The only reason they'd gotten involved with Sugarloaf was another chance cold call: back when the Strosniders living on the top of the mound wanted to sell their home in 2009, the Missouri state senator at the time, Ross Carnahan, had reached out—apparently he had a specific interest in native history.

At the end of our conversation, I asked whether this project had changed Dr. Hunter's idea of art. She paused for a long time. "Not really. I really don't give it a lot of thought."

"At least you didn't have to talk to Joan for a while," I suggested, which made her chuckle. But then she said, "I need to pay Joan a visit." At this point, they've known each other for over 15 years.

Whenever I asked McAnally why he thinks curators and artists should try to do things like rematriate land, he replied with some variation of: "Well, whose job is it?" It's not, he pointed out, like the arts are unbesmirched by histories of displacement and heritage destruction. 16 mounds were flattened to make way for the 1904 World's Fair ("Louisiana Purchase Exposition"), including one of its central venues, the St. Louis Art Museum (which exhibited a new NRO video titled *Give it Back: Stage Theory* during Counterpublic in 2023). By tackling Sugarloaf as a site for exhibition and for transfer, Counterpublic and NRO have pointed out how the history of art and the history of dispossession are often the same story.

American institutions have become accustomed to paying tribute to their own complicity with things like diversity initiatives and land acknowledgements. Over the past decade, in the general "arms race for land acknowledgments," Polys said, these statements have become somewhat rote incantations. The noticeable uptick in exhibitions about social justice might help raise consciousness in certain audiences, but these projects can also pave over the hypocrisies that produce the problems they depict. In 2020 at MoMA in New York, the exhibition *Marking Time: Art in the Age of Mass Incarceration* opened. While the show foregrounded the work of incarcerated artists, protestors pointed out that BlackRock CEO Larry Fink, the second largest shareholder of private prison companies, sat on the museum's board—and called for him to step down. Such are the contradictions inherent to institutions that figure themselves as the terrain of political struggles which stop at their doors. As Adam Khalil put it in one magazine interview: "How do you 'decolonize' an institution? Well, practically speaking, you'd get rid of it."

McAnally told me that "this kind of rising interest in art with a social conscience drives me insane, because most of the time it's almost like using political language to launder money *into* the art world… It's like taking things from community organizing or political organizing and moving it into an art sphere where it becomes symbolic and it loses some of its concrete force." With Counterpublic, he wants to run the code backwards.

Art that works in service of material change is sometimes described as a kind of "cover" or "mask" for the artists' "real" agenda. I heard this reaction a few times during my reporting and found it curious, especially when it came from people who were familiar with the last century of what might broadly be called socially engaged art or relational art.

"If conceptual art and land art can be art, then giving it back can totally be art," Adam said.

"It's not art because it has utility?" Polys asked. "Such a silly idea."

Zack pointed out that they wanted to "instrumentalize the resources that the art world has to offer," but this is not some kind of trick. Nobody is being bamboozled: they're being invited, and artistic integrity is not being denigrated. Polys added: "I love art too much for that. Art can just be the change."

I mentioned that Dr. Hunter didn't seem preoccupied with the art aspect of Sugarloaf's rematriation. "It makes sense that for someone like Dr. Hunter or for many Indigenous people, contemporary art is irrelevant," Polys said. I took this to mean that NRO's intent is not to convince the Osage, or anyone, that art matters. It's to make art, to leverage their industry to push rematriation forward, and in doing so to create new relationships. In what other situation would Dr. Hunter and McAnally and Heckenberg—and I—all be chatting?

I have found it hard to resist turning Heckenberg into a living embodiment of settler colonialism. It makes no sense to pin the blame for centuries of dispossession on one woman; yet rarely do you come face-to-face with someone who so perfectly encapsulates the contradictions and complexities of what it means to *possess*. I asked the three members of NRO how to reconcile my thoughts about her. Polys, who, at 47, is a decade older than the Khalils, asked with a wry smile: "So this is the moment for simultaneous empathy and critique? And maybe a note of hope? I think, Zack, you're going to do that, right?"

Zack smiled kindly. "I don't know if Joan has to be 'reconciled,' so to speak." He went on: "I think she also has a real intense interest to give it back. And I think she has a real intense interest to be at the center of a story too, as much as she seems like she doesn't want to. It's a lot of oscillating back and forth, in a way that's troubling but, to have empathy, understandable. She grew up there, that's the place that she knows as her home." Yet, "for Joan, that's a very personal conversation. It's her legacy. It's her land… It's tied up with all these really intense, traumatic emotions—and such a huge amount of change—having to consider one's own mortality, having to consider one's own legacy."

Polys agreed: "She encapsulates this idea of the personal and the structural, and that can be uncomfortable. And I think within that dynamic, she has a desire for recognition—of her stewardship, holding space and also, her trauma." Although the Osage could not be more explicit that they want the mound to be sovereign tribal territory, at least Heckenberg hasn't sold the land to a trucking company. Heckenberg understands herself to belong on, and to, the land. In Zack's words: "Politically and personally, I think she's also expressed a real desire for indigeneity herself. She really likes Indians, or the idea of Indians."

That reminded me of something that Heckenberg told me in her yard that had unexpectedly moved me. "I've only been on this planet 85 years," she said, pointing to the ground. "My life span is so short compared to how old this is," this being Sugarloaf. While she could have been saying that her choices weren't important in the scheme of things, I got the sense that she was underlining the way that her life is inextricably tied to the ancient mound, and the fact that she, too, is part of its history—which is true.

On the final Sunday of the triennial, a large audience gathered at the base of Sugarloaf to sit on Anita Fields's platforms and watch a performance by the Wah.Zha.Zhe Puppet Theater, a troupe organized by Fields's daughter, Welana, and her son, Nokosee, among other Osage community members invited by Puleo. People of all ages maneuvered huge papier-mâché elk and birds in a retelling

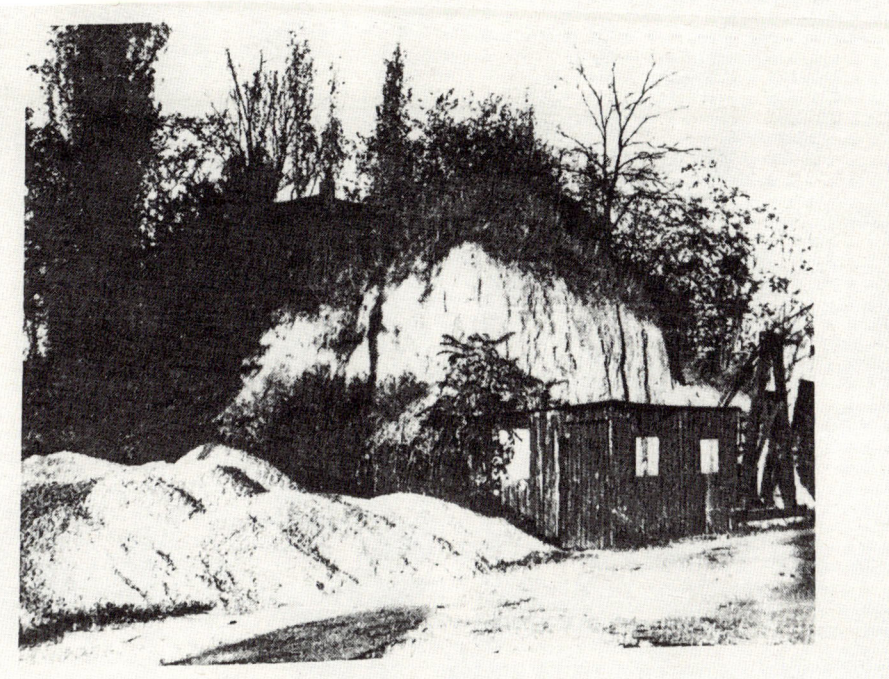

Fig. 2—View of Sugar Loaf Mound in 1940, south face. Photo from the St. Louis Post-Dispatch, October 13.

Photo copied from Illinois assn. of archeology journal of 1982 (see pg 18)

of the Osage origin story. They played a recording of an elder speaking the Osage language, which the UNESCO World Atlas of Languages says is "not in use." The heat and humidity were so oppressive that my dress was soaked with sweat. Several performers and audience members were visibly emotional. Heckenberg stopped by briefly on her way to church.

Counterpublic's project statement calls it an exhibition working "towards generational change." The Mellon Foundation grant to rematriate Sugarloaf expires after two years, in September 2024, so there's a set timeline. McAnally, however, has said that when it comes to Sugarloaf, "There are no deadlines for this work. It's human relationships and legal processes… We as an organization are committing long-term, until there are no further paths or until it's done."

Rematriation is not about essentializing native connections with the land. It is, Adam pointed out, not even always about the land—it's just that land is the most "legible" way to conceive of sovereignty. The long-term relationships enabled and required by land transfer projects are what allow sovereignty to be constructed in new ways, with new accomplices. Emphasizing the slowness of this work, Zack said, "You have to move at the speed of trust. That's been our relationship with Joan. Just having to show up over and over again. Call over and over again."

And so, after a long pause, in the fall of 2023, McAnally and Dr. Hunter visited Heckenberg again for a chat, and by December, she had verbally agreed to a purchase option that would allow her to leave whenever she wants. This time she has agreed "with certainty," reaffirming her decision twice. Counterpublic's legal representative drafted an agreement that accords with the Mellon grant stipulations, and as of early 2024, the document was with the Osage, who are reviewing it closely before all parties present it to Heckenberg for a signature. The offer to purchase the Kappa Psi land still stands, should the fraternity choose to take them up on it.

"It might not feel like that at first, but there's something to be gained for everybody by giving it back," Zack told me. "It would be hard to deny that our country, our nation, the world, maybe, is on a path that's fundamentally unsustainable. I'm not saying that native people are superhuman new age forest bunnies that are going to solve global warming in an instant. But there's a reason we're in this position right now."

When he said this, I remembered Heckenberg's odd use of verb tenses: what *would* happen or what *will* happen, with no mention of how things would get there. I realized that sometimes NRO does the opposite: they speak about the future in the present tense. As if it were already here. As if it were happening. Because, as Polys told me, "belief shapes reality." Or, in Zack's words, "the rematriation of all Indigenous land and life" is a slow process, but—here he grinned—"It'll be reality before you know it." ●

SWAMP DOGG

If You Can Kill It

I Can Cook It

Before Snoop Dogg, there was Swamp Dogg. For the past 60 years, this legendary singer, songwriter, and record producer has shaped the history of soul, country, hip-hop, and R&B with his singular voice and ideas—including the rather audacious plan he hatched in 1972 to create a cookbook to rival Julia Child's. For decades, *If You Can Kill It I Can Cook It* was known only to Swamp Dogg's friends. But now his culinary opus, from which a few recipes are excerpted here, is being released by Pioneer Works Press. Swamp's tome is more than a compendium of comfort food classics; it includes archival photos and stories to go with dishes he's crafted, over the years, to honor the artists and people he loves. *If You Can Kill It I Can Cook It* is a glimpse into the savory life of a cultural genius.

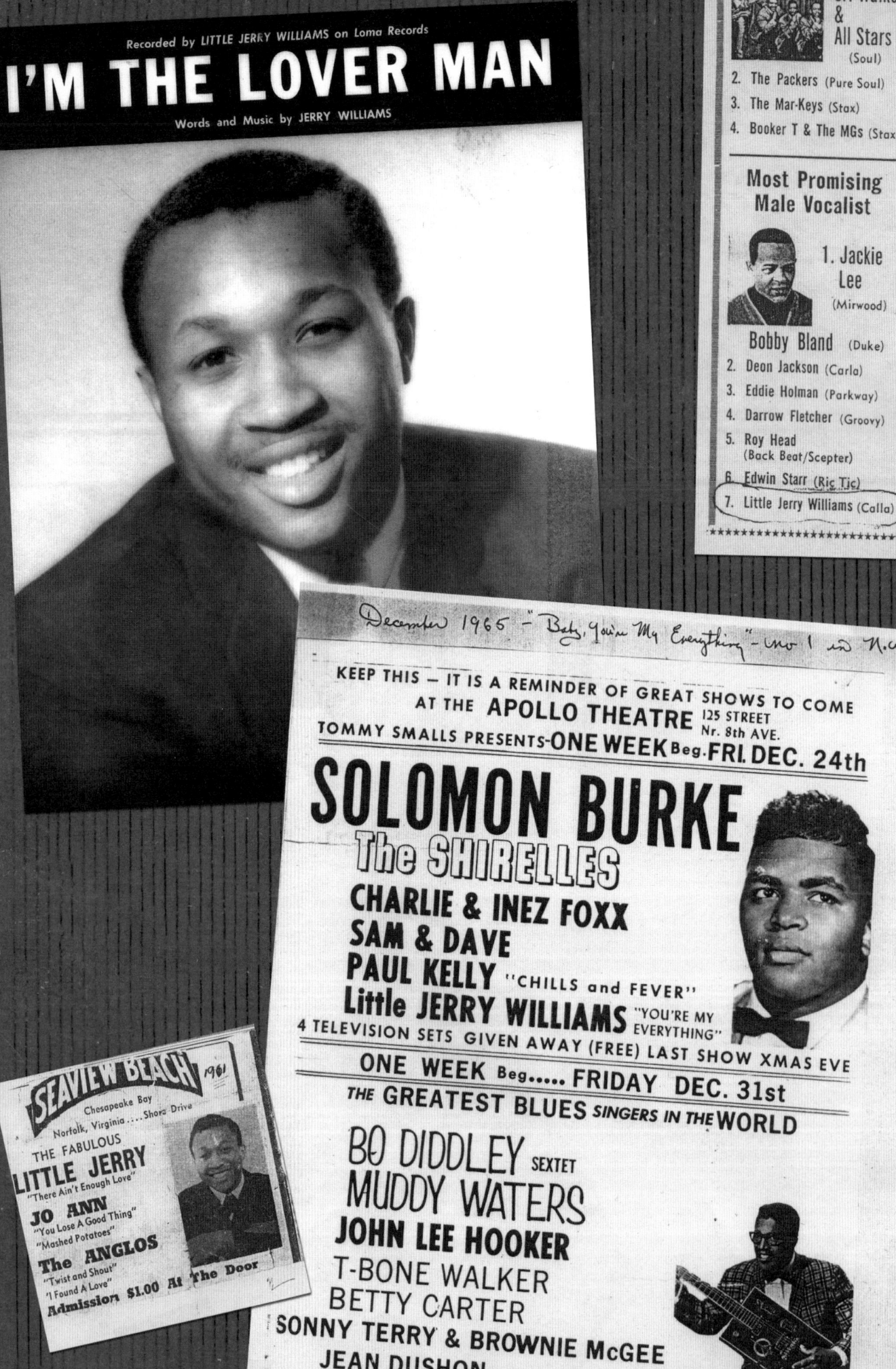

BAKED-BEANS
BO DIDDLEY

Bo Diddley has always mesmerized and amazed me with his weird-ass songs, weird looking guitars, and Africa-meets-Chicago musical grooves. Bo used to always play the Capitol Theatre in Portsmouth, Virginia. with his real weird looking maracas player, Jerome Greene, who reminded me of one flying over the cuckoo.

Bo introduced fun into rhythm 'n' blues, which is why I dedicated a dish to him that's usually prevalent at fun filled events and holidays.

I approached Bo about eight years ago regarding me producing him and placing him with a label. After listening to his twenty-minute retort on his greatness, his sheriffing in New Mexico, and who the fuck am I, I felt like something he'd eliminated from his ass. One day I'll learn how to be a devoted fan and stay the hell out of harm's way.

- *1 32 oz. can pork 'n' beans*
- *4 tbsp. butter*
- *½ cup brown sugar (old fashioned or light brown)*
- *2 tbsp. flour (browned)*
- *4 tbsp. water*
- *1 tbsp. grated orange peel*
- *3 tbsp. dark heavy corn syrup*

- Melt butter over low heat in a heavy frying pan. Add brown sugar to melted butter and cook for 5 minutes or until butter and sugar become thick and sticky. Combine flour and water with pork 'n' beans. Add beans to fry pan mixture and stir. Sprinkle orange peel over beans. Remove from stove and place in 350 degree oven for ½ hour uncovered. Pour syrup over beans and return to oven for 10 minutes.

- Remove and cool for 10 minutes.

(Serves 8)

LIONEL (HAM)PTON

In 1961, I replaced Pinocchio James as the singer for Lionel's big band at Pinocchio's request, while he recuperated from the flu. Here I was, twenty years old, in Washington, D.C., for the opening of the WOOK TV station, standing beside Lionel's vibraphone and singing my black ass off. ''Honey Hush,'' ''Flip Flop and Fly,'' and ''Since I Fell For You''... I'm shouting the blues and his band is performing a colectomy on me and I got paid. I would have paid him. A singer has not lived unless he or she has had the opportunity to stand in front of a Lionel Hampton or Duke Ellington, or Count Basie, or Woody Herman or their equivalent and belt their little hearts out.

- *1 6 ½ lb. smoked ham*
- *whole cloves to taste*
- *2 12oz. can Coca-Cola*
- *½ cup brown sugar*
- *½ tsp. dry mustard*
- *2 tbsp. orange juice*
- *1 scoop flour*

- Score ham in diamond patterns about ⅛-inch deep. Push cloves into each corner of the diamond. Place ham in a cooking bag with a scoop of flour. Pour the Coca-Cola over the ham, tie and then close cooking bag, place bag in a roasting pan, put pan in a 350 degrees oven for 1 hour. Remove ham from oven, open bag and baste, put the ham back for another ½ hour. Take ham out again, mix together the brown sugar, dry mustard, and orange juice, spoon over the ham, tie bag back together, and put back in oven for ½ hour more.

(Serves 12-15 hearty appetites.)

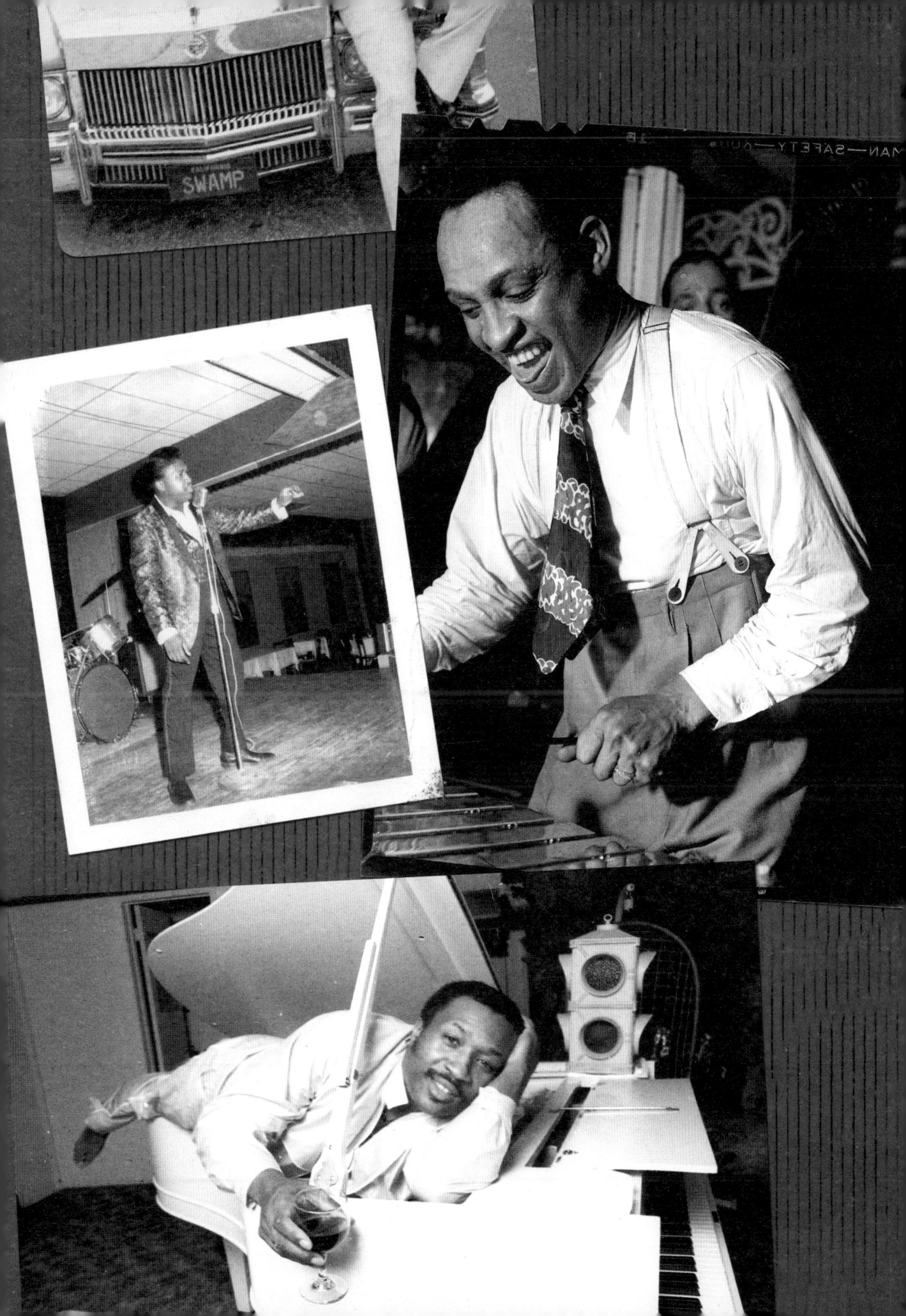

1650 Broadway, Suite 1211, New York, N.Y. 10019 Tel: 757-6921

BOTANIC
RECORDS INC.

JERRY WILLIAMS, JR.
A & R Director

1-866-DOGG-FUD

GEORGE JONES POTATO PLATTER

George Jones is one of a kind. He earned the nickname ''No-Show Jones'' by not showing up for performances, and his fans always forgive him and buy tickets for his next scheduled appearance. George has gotten drunk and driven his car into trees, embankments, and you name it, totaling his auto but walking away, every time, unscathed. George has admitted to consuming enough alcohol to kill all of Hell's Angels and I don't think he has an ulcer. George is unstoppable, immortal and the greatest country artist in the world. His records have sold in the tens of millions and I defy anybody to open their minds and listen to some music by the Jones boy and not be captivated by his soul and sincerity.

I met George when he was recording for Musicor in the '60s and that was a memorable event for me.

George, this is my thank you for all of the wonderful music you've made for me to enjoy. Who's gonna fill your shoes?

- *4 large white potatoes*
- *1 cup milk*
- *1 cup water*
- *¼ cup canola oil*
- *1 10 oz. can condensed cream of celery soup*
- *1 onion*
- *2 tbsp. brandy*
- *¼ tsp. pepper*
- *½ tsp. paprika*

- Heat oil in frying pan. Peel potatoes and cut into ¼-inch slices. Put potato slices in hot oil, add pepper, cover and fry for 10 minutes over a medium heat. Peel onion and slice into ¼-inch rings and pull apart. Turn potatoes and add onion and brandy, cover and simmer for 15 minutes, turning often to ensure even cooking. Add soup, water, milk, and paprika and stir gently. Cover and simmer for 30 minutes, stirring often so mixture doesn't stick.

- Great with turkey, baked or fried chicken.

- Serve hot.

(Serves 6)

YVONNE'S TOMATO PUDDING 33 1/3 RPM

My wife, manager, lover, buddy, and confidant, Yvonne, loved this dish and would kill for it. There were times when I'd make it twice in one week for her and she still remained slender, tender and tall. Libby taught me this one also but she learned it from my Aunt Kini, her mother.

This fantastic dish has to be done in two steps, but the end results will make every step worth its weight in calories.

TRACK I (8 CRUSTY BISCUITS)

- *2 cups all-purpose flour*
- *2 tbsp. vegetable shortening*
- *¾ cup cold water (do not substitute milk)*
- *1 tsp. baking powder*

- Preheat oven to 450 degrees

- Combine flour, baking powder, and shortening. Mix together thoroughly with spoon or hands. Pour in water gradually while working pastry together. When flour is soft and light, not sticky, turn out on floured board. With floured hands pat dough until smooth. Roll into an oblong and cut out 8 biscuits. Place on ungreased baking sheet, 1 inch apart and bake for 15 minutes.

- No do not use canned biscuits under any circumstances.

TRACK II (PUDDING)

- *1 16 oz. can peeled tomatoes*
- *1 cup sugar*
- *8 oz. butter (melted)*
- *1 tsp. nutmeg*
- *8 large biscuits*

- Put tomatoes in a bowl and mash with a potato masher. Add butter, sugar, and nutmeg, stir until blended. After the biscuits have baked remove from oven and reduce heat to 400 degrees. Wait 5 minutes and put tomato mixture in the oven for 30 minutes, in a 2-quart baking dish. Remove the tomato mixture from the oven and add biscuits crumbled up in tomato mixture. Stir and place in oven and bake until biscuits are golden brown.

- This is another great dessert to serve at those times when you are bored with the norm.

(Serves 6)

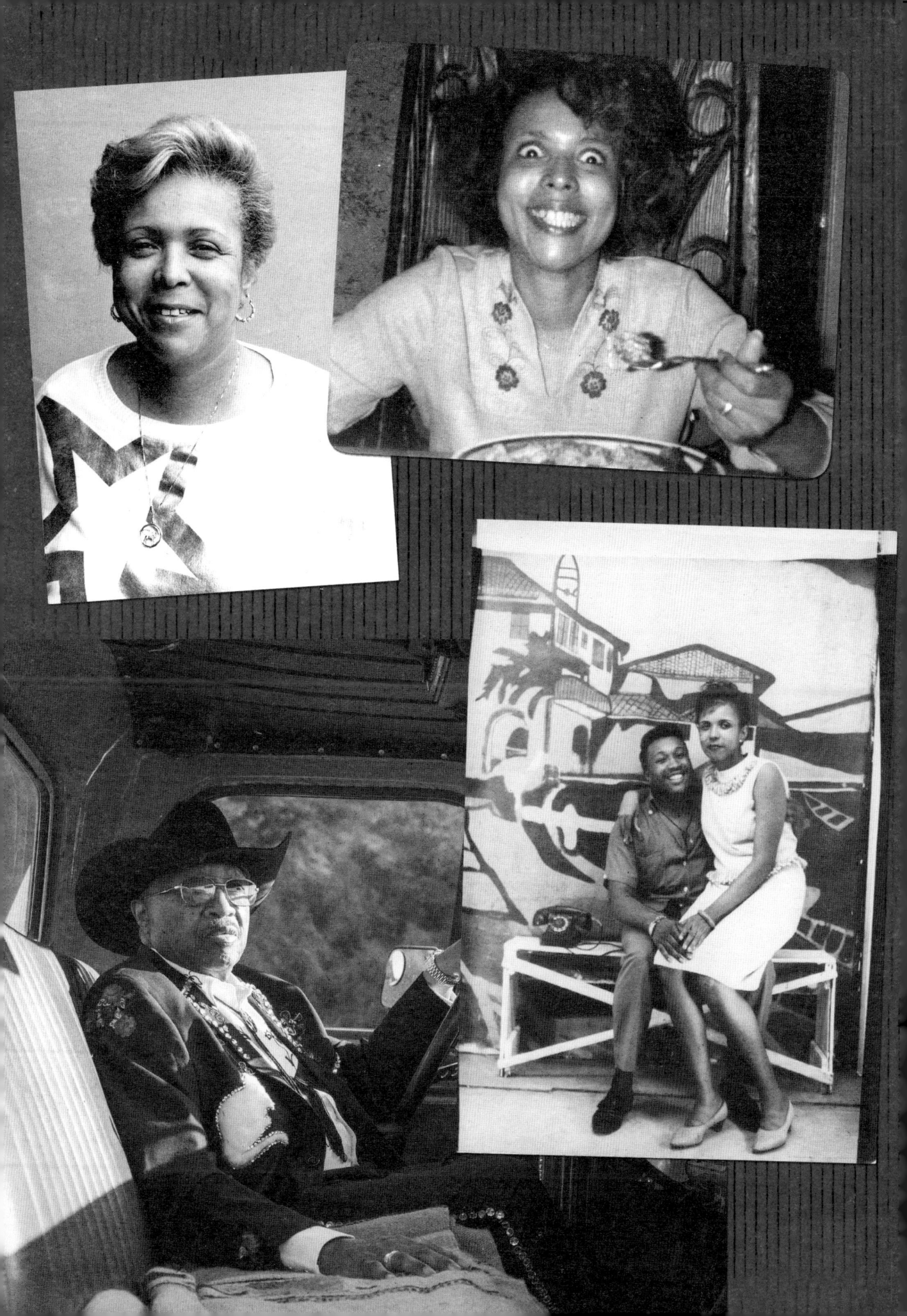

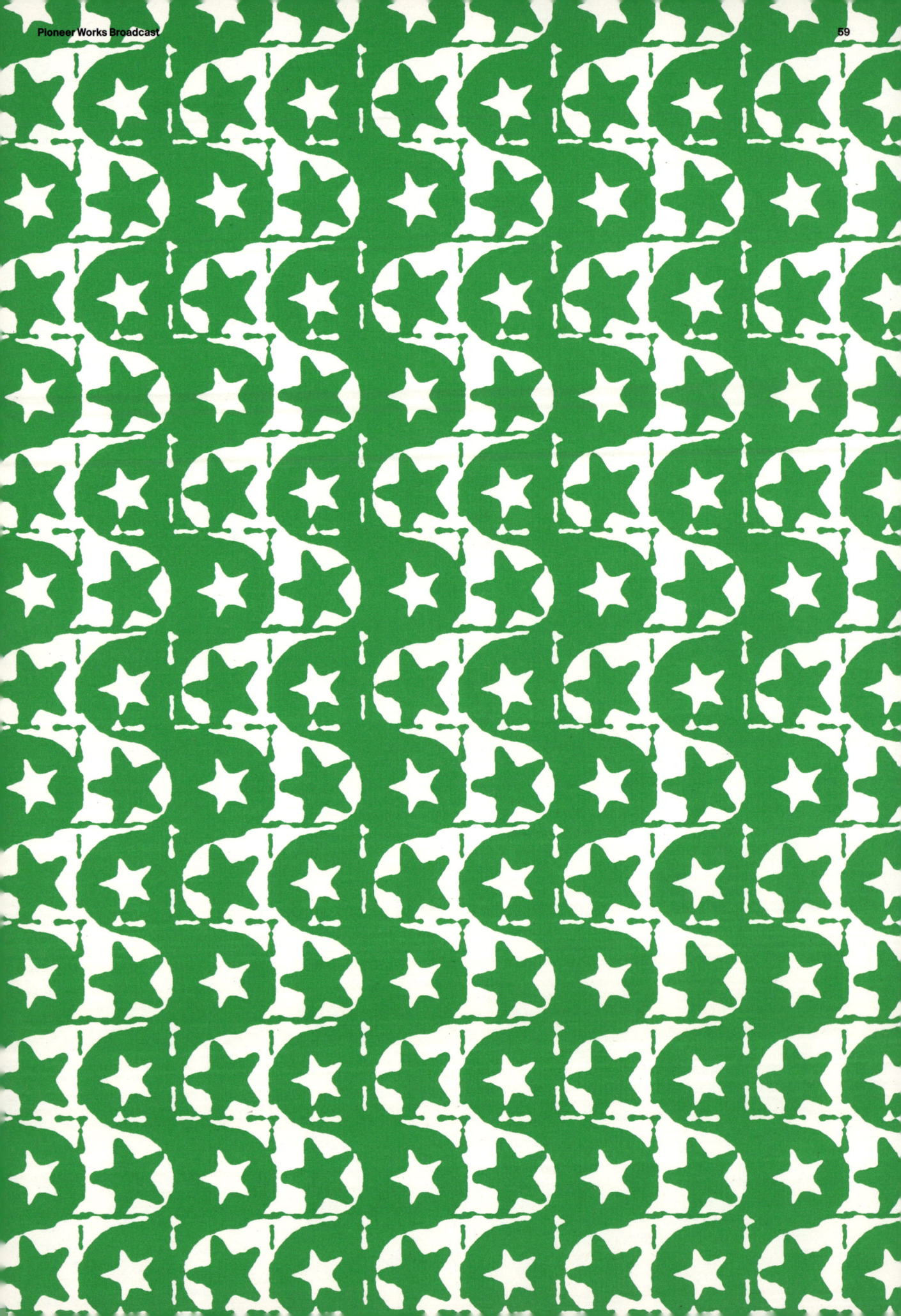

THE GREAT PSYCHEDELIC EXPERIMENT

NATASHA BOYD

Researchers mined an old drug forum and fed the entries to an AI. The result could augur a new class of psychedelic-based antidepressants.

"If there is one quick truism about psychedelic drugs it is that anyone who tries to write about them without first-hand experience is a fool and a fraud."

—*Hunter S. Thompson*

People have been doing psychedelics more or less intentionally for at least 3,000 years, and yet our understanding of how these substances interact with our nervous systems is still in its infancy. Until recently, illegality hampered clinical research. But even now, as some psychedelics pass the threshold to decriminalization, large-scale trials remain few and far between—a problem for research into such a variable experience. That's not to mention the inherent contradiction of conducting clinical trials on any social (or at least highly contextual) drug. People usually trip in scenic environments with their friends. That's hard to replicate under scientific observation.

The temperamental nature of psychedelics is essential to their appeal. Drug nerds—the true bearers of this occult knowledge—love to share their unique experiences with each other. To the researcher, this same variability represents a bottleneck, and a costly one. Venture capitalists are pouring millions of dollars into new psychedelic-based antidepressants that move beyond the classic SSRI model, but without a precise index of the neural processes at work, pharmaceutical patents are unlikely to be approved.

Last year, however, a group of interdisciplinary researchers announced a simple but powerful workaround. Using AI and brain imaging, they found a way to draw directly from the experiences of some of the internet's chattiest psychonauts—a small breakthrough for psychedelic research that could inspire a new class of hybrid drugs.

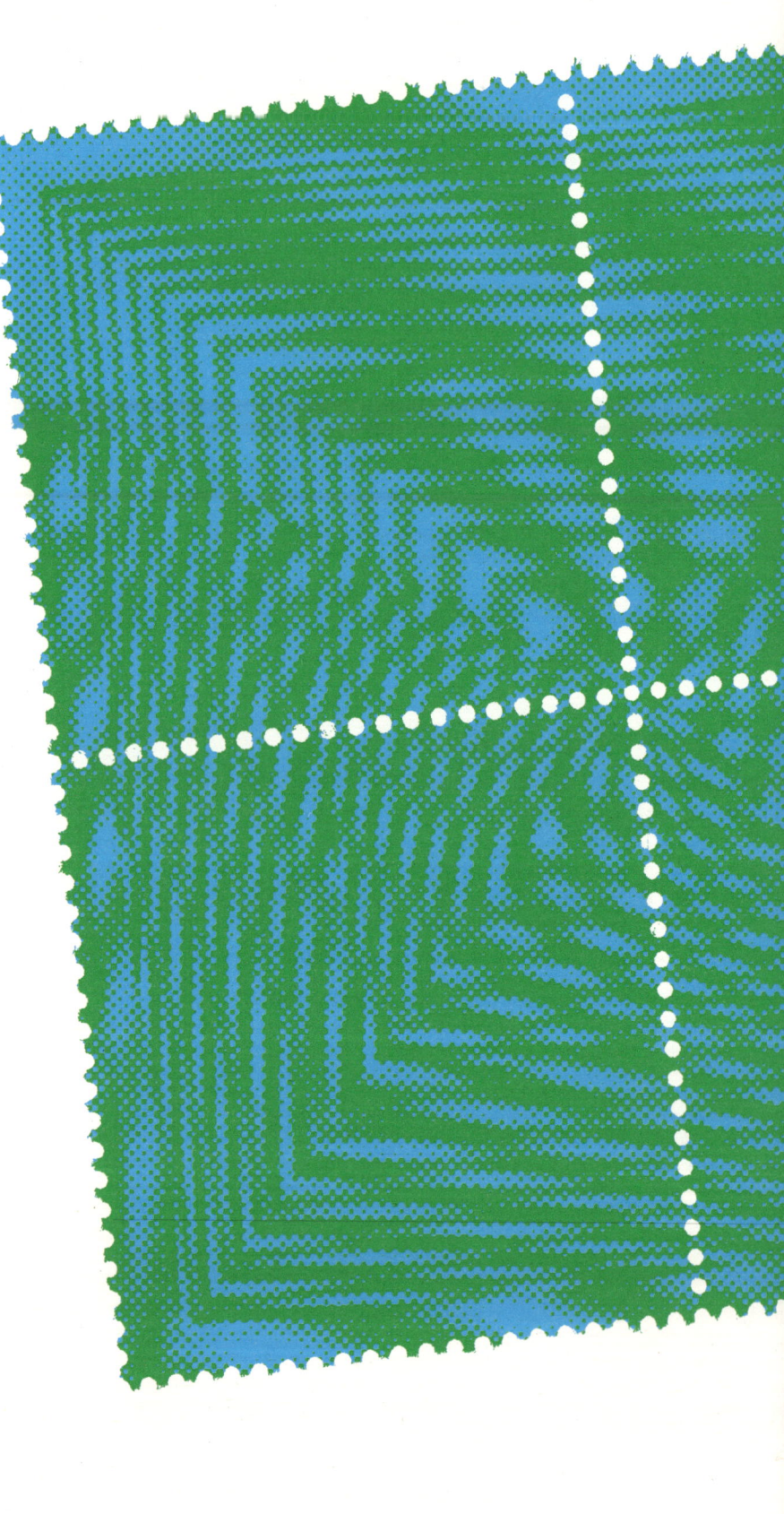

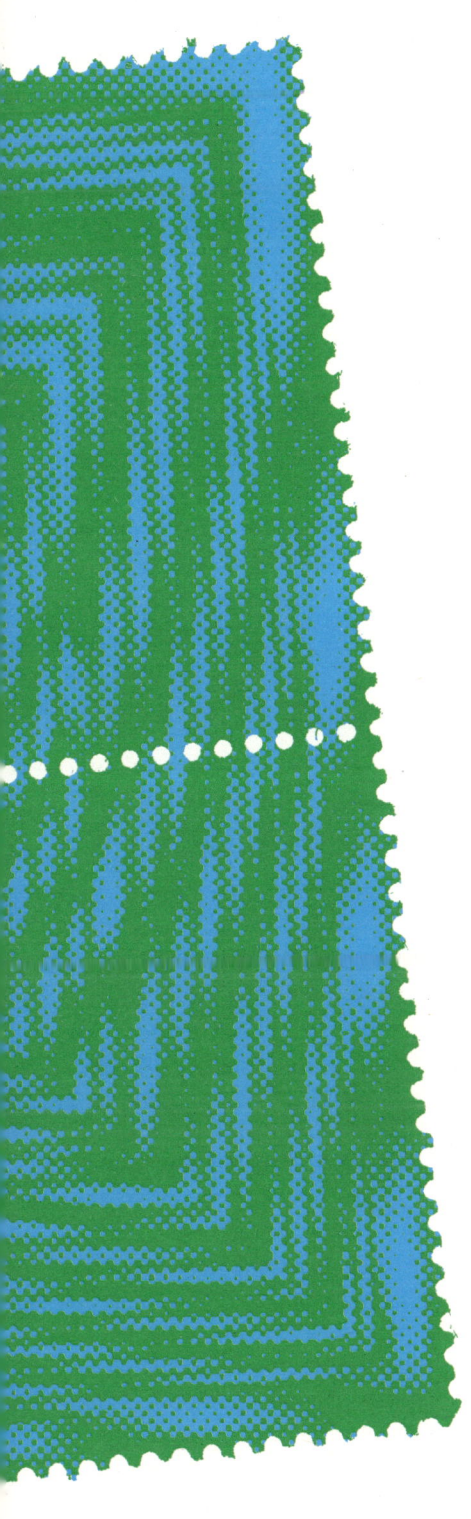

Samuel Friedman, a machine learning researcher at MIT, came to the psychedelic question via an enigmatic region of the central nervous system called the "Default Mode Network." Formerly known as the "task negative network," the DMN is thought to govern some of our most introspective behaviors. Neuroimaging shows increased blood flow towards this area during daydreaming, reminiscing, "wakeful rest," and the deep contemplation of our relationships. An overactive DMN is also associated with a range of mental health issues—particularly, the obsessive pondering of one's own unhappiness, known as "rumination," a symptom of major depression. Psychedelic treatments have shown great promise at interrupting an overactive DMN, apparently correlating with subsequent improvements to a patient's mental health.

Hoping to illuminate these connections, Friedman organized a talk in 2019 with Danilo Bzdok of McGill University, who was working on a comprehensive mapping of the DMN, and his childhood friend Dr. Galen Ballentine, a SUNY psychiatrist pursuing research into the therapeutic potential of hallucinogenic drugs. Over dinner, afterwards, the three men lamented the miniscule sample sizes of most studies into psychedelic-assisted therapy. Drug trials of this sort have particularly stringent exclusion criteria, resulting in volunteer disqualification ranging from 90% to 96.3%. Notably, people already taking antidepressants aren't eligible. Bzdok wondered if there wasn't some pre-existing data set out there that could be analyzed. That's when Ballentine remembered the Erowid Center—a labyrinthine online drug forum, where thousands of people have been sharing their experiences with psychoactive substances for over two decades.

This opening yielded an experiment combining all their specializations. In the following months, the team downloaded nearly 7,000 testimonials for 27 different psychedelics from Erowid's archives. They used AI to boil down commonly recounted experiences with a particular drug into their most essential, expressive keywords. Bzdok's brain imaging maps subsequently connected these keywords to a particular psychedelic as well as a specific region of the nervous system.

The fruit of this labor is "Trips and Neurotransmitters: Discovering Principled Patterns Across 6850 Hallucinogenic Experiences," the largest ever study of psychedelic influence on the brain. Published in March 2022, it's been hailed as a "Rosetta Stone" that translates symptoms to molecules. Along eight distinct axes of psychedelic experience, it offers a precise and yet expansive map of neurotransmitter receptor combinations that need to be stimulated to induce a specific state of conscious experience. It also clarifies how the DMN ordinarily works to maintain a stable sense of self capable of withstanding the constant stimulation of being in the world—and how psychedelics might shake it up.

Obtaining this volume of actionable results through a classic clinical trial would have required a nearly prohibitive amount of time and money. "One of the mysteries about these [machine learning] models is that they need so much more data than humans to learn things," Friedman tells me over Zoom. "But if you have the data, then they are able to learn things that humans can't."

On a shared screen, he talks me through a presentation of their findings. The slide in front of me has two columns of words, one in red and one in blue, flanked by scans of the human brain, with clouds of color indicating neurotransmitter activation. The column in red has words like "love," "dancing," "magic," "happy," "crying," and "time," while the blue column contains "distortion,"

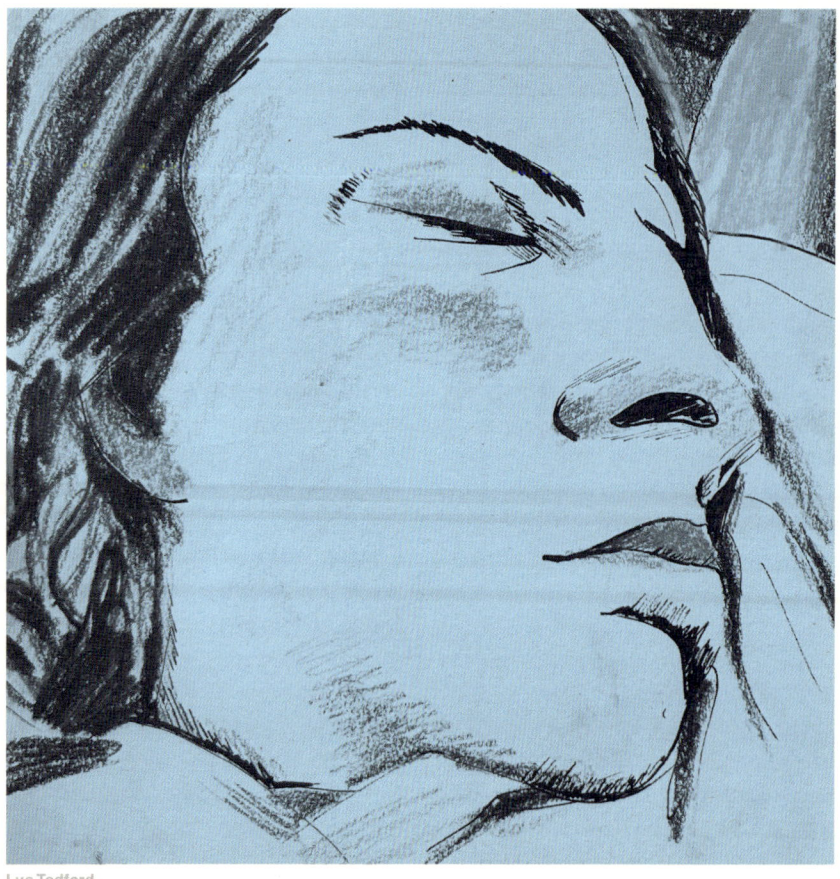

Luc Tedford

"deepening," "nausea," and "voices." Even reduced by the AI model, the remaining kernels of language retain the vividness of all Erowid entries.

Erowid was founded in the midst of Bill Clinton's presidency, when DARE officers were still giving lectures nationwide on the "Three R's": Recognize, Resist, Report. In opposition to this abstinence-focused approach, the founders of Erowid—a Bay Area couple known only by the sobriquets "Earth Erowid" and "Fire Erowid"—created an open-access educational resource. Their stated goal was a "world where people treat psychoactives with respect and awareness; where people work together to collect and share knowledge in ways that strengthen their understanding of themselves, and provide insight into the complex choices faced by individuals and societies alike." It remains ad-free and looks much the same as most websites from the earlier and more hopeful days of the internet.

Each drug on Erowid has its own "experience vault," full of first-hand testimonials that have been vetted for authenticity and accuracy. Beyond that they're pretty free-form. "At about 8 PM, I chose to vomit (as I was fairly nauseous and thought it might expel some of the toxins). During my toilet-filling episode, I thought I was going to die and I asked God to help me," goes one testimonial on psilocybin. "With each spew, I felt better and I got the message that Jesus had saved me." Another recounts the pleasure of "rubbing my hands and body against the bark of some large evergreens."

While some entries can be bleak—particularly for harder drugs like meth or heroin—the vast majority are written in a companionable, curious voice that will be familiar to anyone with an older sibling or cousin who likes to test the limits of consciousness from their own backyard. The testimonials include highly specific descriptions not just of the chosen amount and imbibing method, but also the subtle shadings of each experience; sometimes with humor, but always with rigor, vibrancy, and clarity, often down to the passing minutes. These are good faith arbiters, truly interested in exploring the variance of human perception and making sure others could do so safely. There are none of Hunter S. Thompson's "fools or frauds" here, though any one writer tends to give the distinct impression of being a bit of a weirdo.

Erowid contributors would probably be intensely curious about the new substances their testimonials might be used to formulate; how their words, written in an online forum as much as 20 years ago, were fed into an artificial intelligence program and then translated into molecules.

Their experiences could help bridge a tricky divide. "The problem is not the drug—drugs are just molecular tools—but rather, not pairing the right tool with the right patient," noted two notable psychiatrists, summarizing the implications of the study. "The task facing the companies and researchers developing psychedelic compounds is to identify which specific compounds alleviate which specific types of human suffering."

This shortcut could speed up the effort to patent new psychedelically derived drugs—which could save a lot of lives. "The U.S. is going through a profound mental health crisis that has been exacerbated by the COVID-19 pandemic," Bzdok reflected when the paper first dropped. "Yet there have been no truly new psychiatric drug treatments since Prozac […] These insights may lead to new ways to combine existing or yet to be discovered compounds to produce desired treatment effects for a range of psychiatric conditions."

In for-profit pharmaceutical research, the rush to save lives and the rush to make billions are perfectly complementary. If an entire nation is experiencing a mental health emergency, it's worth considering whether this is something that can be individually remedied by the right combination of molecules. Depression diagnoses increased by more than 300% after the advent of SSRIs. A new psychedelic-style Prozac would be a gold mine for any company lucky enough to patent it and then prescribe it to a population in crisis.

In recent years, for-profit companies like atai Life Sciences, MindMed, and Compass Pathways are flooding universities with millions of dollars in support of psychedelic research. MindMed, for instance, provided enough funding to NYU to support four research positions over five years. Their efforts have been buoyed by loosening regulations, including the FDA's approval of ketamine-derived nasal spray as a treatment for major depression. Canada set a precedent last August when four terminally ill patients were granted access to psilocybin as part of end-of-life care. In the United States, meanwhile, voters have passed measures legalizing shrooms for controlled therapeutic use in Oregon and decriminalizing plant-based psychedelic substances in DC. Legalization, however, does not seem to be a major concern for these companies, beyond how it affects their ability to conduct research.

One of the giants of this emergent field, atai founder Christian Angermayer, said recently that he supports decriminalization but thinks legalizing psychedelics could create a backlash for the industry. "Biotech is all about having a monopoly," he said, in an interview with *Insider*. "That's the whole of biotech. You do something novel, you finance it, and you own it."

Getting on the right medication can be life changing, but for many it's an arduous process. Psychiatrists—or, for that matter, neurologists—can't reliably predict who will improve on antidepressants and who will not. Until recently, clinical professionals operated on the assumption that depressed people had a serotonin deficiency and that SSRIs worked by boosting it; however, a paper published last July presented evidence that people with depression have the same amount of serotonin as people who are not depressed. The updated consensus is that antidepressants can help the brain form healthy connections between cells that have previously withered, possibly due to stress, and that serotonin might simply be acting as a growth factor. Part of the excitement around psychedelics is that they demonstrate the ability to operate in a similar fashion, improving mental health by rapidly inducing neural plasticity. A recent study at Yale University showed that mice given one dose of psilocybin grow longer and thicker dendrites—the tapering branches that extend from the body of a neuron—yielding a full 10% increase in neural connections.

This kind of increased neural plasticity is associated with a concurrent loosening of the DMN's top-down control over what constitutes the inner self, leading to a leaky filtering of sensations, thoughts, and feelings—essentially, a disintegration of normal self-awareness. Regions of the brain that aren't usually in conversation end up talking to each other. Psychedelics seem to have the special aptitude to slacken certain connections while strengthening others. Crucially, the more powerful this dissolution of ordinary processing and ego-maintenance, the better the clinical result in the long term.

For people with treatment-resistant depression—that is, patients who do not improve with medication and psychotherapy—the acuteness of this oft-described "mystical" experience turns out to be particularly consequential. Compass Pathways, a London-based startup, found that three weeks after psilocybin-assisted therapy, 29% of people with treatment-resistant depression who received the highest doses were in remission. In trial scenarios, such high-dose applications lead to "spiritual," "mystical," or "religious" experiences. A 2018 study on hallucinogens notes that, "regardless of the terms chosen to define them, evidence suggests that profound psychological experiences can be predictive of subsequent psychological health," whether induced by psychedelics or not.

Despite its obvious import, research scientists remain wary of the mystical experience. For one, it's challenging to quantify. The same 2018 paper notes that "mystical" suggests "an association with the supernatural that may be obstructive or even antithetical to scientific method and progress," and that they are not keen to "endorse any associations between it and supernatural or metaphysical ideas" by describing the phenomenon. Researchers have done their best to get around this by developing various questionnaires for people undertaking a psychedelics trial. The Mystical Experience Questionnaire (MEC), for instance, features questions designed to quantifiably measure a subject's "transcendence of time and space"; "sense of awesomeness, reverence, and wonder"; and feelings of "ineffability and paradoxicality."

One of the most important facets of "Trips and Neurotransmitters" is that it highlights this key aspect of psychedelic-based therapy on a molecular level. "Earlier research on ego dissolution has placed a focus on binding to the 5-HT2A receptor," a single kind of serotonin receptor, the authors write, but "together, a wider group of receptors may individually or collectively underpin the process of self-disintegration that is thought to be critical for the success of hallucinogen-assisted psychotherapy."

There's another major obstacle facing hopeful pharmacologists. Unless the hallucinatory effects are eliminated, prescription medication derived from psilocybin, MDMA, and LSD can't be used by patients without supervision. This would

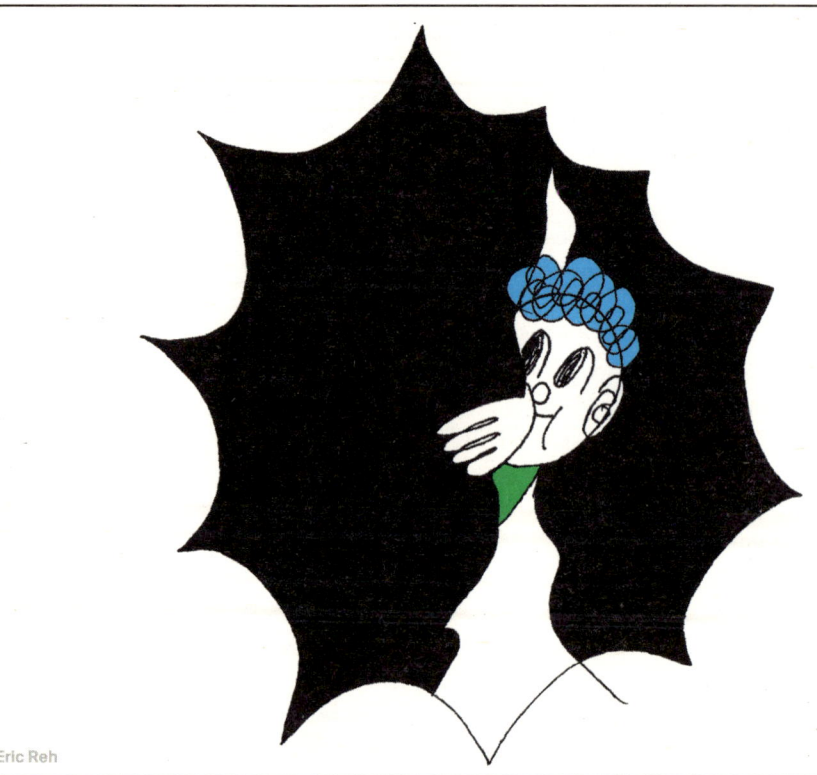

Eric Reh

make it a lot harder to market them as a replacement for the classic SSRI. For that reason, some researchers are attempting to develop a psychedelic analog to CBD, a derivative of cannabis that lacks its psychoactive properties—and so evade stringent legal oversight.

The evidence presented in "Trips and Neurotransmitters" suggests that this will be very hard—the mystical experience and the hallucinatory effects are deeply entangled. "We don't really get to say, 'Okay I want to segregate the visual hallucinations from the mystical experience, and I want it to show me how to break down the experience upon those axes,'" Friedman says. "What did come out [of the study] is there seems to be this very hallucinatory thematic side, with words like 'nausea' and 'sleep' and 'stomach,' and also 'hallucinations,' 'euphoria,' 'visuals,' 'headache.' Then on the other side was this very mystical, intense emphasis on 'reality,' 'breakthrough,' and 'consciousness.' You can imagine pushing along that axis like a knob that you could turn." He smiles. "That would be cool."

How close are we then to actually seeing new drugs show up on prescriptions? Friedman considers for a second, then replies that we might need to start thinking differently about how we conceive of medication. "It's such a different model than antidepressants, where you take it chronically," he explains. "These are one or two interventions with really long-lasting effects. And it seems that the setting matters so much, and the person administering it plays such a critical role…"

Friedman is the first to say that we're still in the "very early days" of viable psychedelic medication, even if it becomes possible to parse out the visual hallucinations molecularly. The mystical-leaning axis of molecule-to-experience in "Trips and Neurotransmitters" is as characterized by intense fear and dysphoria ("terror," "horrible," "death," "fear," "panic," "pressure," "shit," "insane," and "surrender" all appear in this word constellation) as it is by a parallel experience of transcendence ("forever," "lightbulb," "bliss," "void," and "life").

The challenge is philosophical as well as chemical: can we really separate the agony from the ecstasy? ●

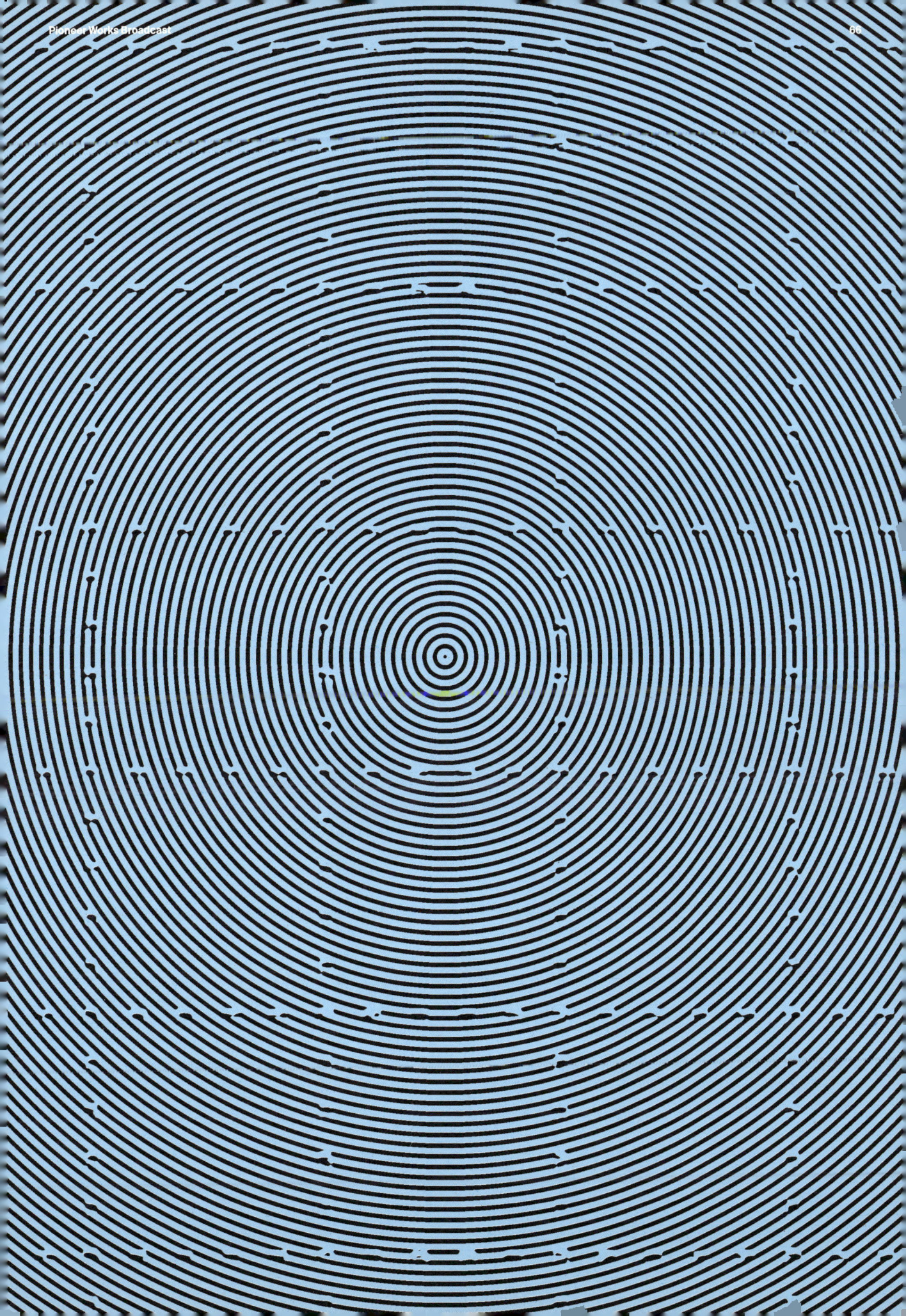

In an epic tale of two islands, Vivien Goldman returns to where she came of musical age—and seeks the source of the songs, shorn of bass and obsessed with "bad mind," vibrating the Caribbean now.

VIVIEN GOLDMAN

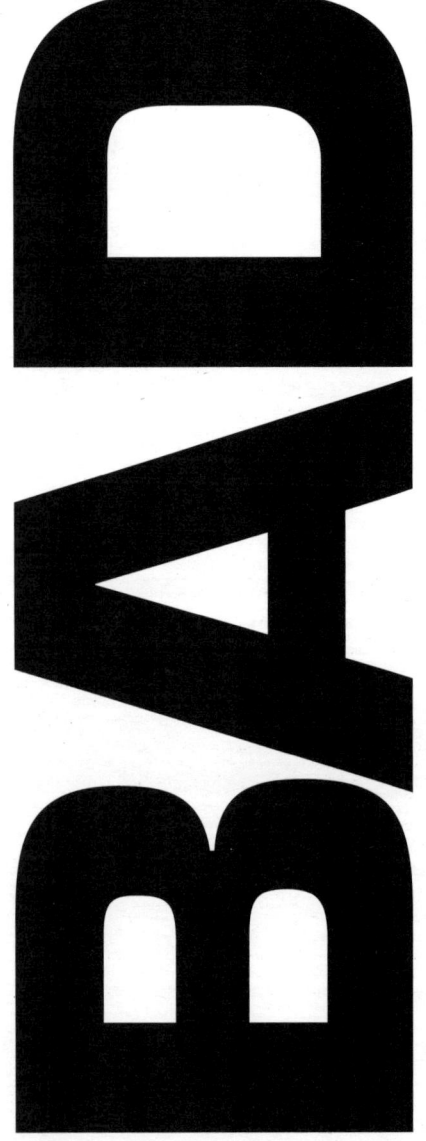

1. BASS(LESS) CULTURE

Perhaps it is in the nature of cultural commentators to glorify or even romanticize the youthful period in which they first found their voice. Thus it is that when I found myself unexpectedly living in Jamaica, a COVID refugee at the start of 2020, I was shocked as an old school reggae lover by the change I heard in Jamaican music. What, no bass?

When I first landed in Jamaica from London in the mid '70s, I was in thrall to dub and classic "one drop" reggae: the music's low end was a fetish. Rounded and resonant, reggae bass was amplified by "sound systems" in the street to roaring melodic thunder; speaker stacks were designed just for it. The instrument was defined as the soul of global reggae culture, and that of Black Britain in particular, when the great dub bard Linton Kwesi Johnson described what bound a generation as *Bass Culture.*

Coming from leafy Hampstead, I had lucked into living in London's Bass Culture country, in loucher Ladbroke Grove. Known for London Carnival—the biggest street party in Europe—it was also home to a number of all-night reggae dance parties in firetrap abandoned properties called shebeens, which provided me with a great social life for a long time. In addition, I found myself briefly working as the publicist for Bob Marley and the Wailers, and then covering Bob as a music journalist, learning from his commitment, compassion, and ferocious work ethic. He became a mentor and set a high bar for the connection between music and activism— his message always powered by that propulsive bass—in Bob's case, coming courtesy of Aston "Family Man" Barrett. Fams had mentored

Robbie Shakespeare, known for his work with Sly and Robbie, Jamaica's session rhythm powerhouse. But both "bassies" have recently died—with Fams's passing last winter, I felt the era of bass-heavy "conscious" reggae fading further into the past.

When reggae was my religion, during the 1970s, I became part of a generation that benefited from a sort of reverse cultural colonialism which followed the colonies gaining independence. The sounds of the Empire's former subjects ruled us all—a process furthered right now by the global dominance of Nigerian Afrobeats. But back when I was staying at Hope Road, reggae's international market was exploding. Famously, oil-flush Nigerian customers would turn up at Virgin Records's offices in London with suitcases of cash to buy up wax of U Roy or The Mighty Diamonds. Pivotal was the ascent of Bob Marley.

In Jamaica in 1976 for Virgin's Front Line label, I overstayed my hotel tab and Bob kindly invited me to stay at his home in uptown Kingston. An ample mini-great house of the late colonial era, the communal space on Hope Road—now the Bob Marley Museum—offered the dynamic theater of people constantly dropping in from all classes and social circles—uptown, downtown, and those like myself, "from foreign." Marley had spent his own boyhood in one of downtown Kingston's impoverished ghettos; and was now determined, he told me, to "bring the ghetto uptown." He continued deliberately, "You have to show people some improvement. Not necessarily materially, but in freedom of thinking." The day after I left for London, gunmen burst in and tried to kill him, wounding two others and leaving a bullet in Bob's arm that stayed there until his death. I'll never forget arriving to the rock paper where I worked, the morning after I flew home, and seeing the shocking news on the telex: "reggae singer shot"—in the bustling commune I had just left.

Now, the period leading up to and following this assassination attempt—a period when Bob fled Kingston for London, and I had a front-row seat for the recording of his great 1977 album *Exodus*—is the subject of a big-budget movie. The worldwide release of *One Love*, the Marley biopic that did big business at the box office in early 2024, is another reason my mind has been returning to those formative days, and contemplating how the new sounds of Jamaica's streets resonate with these even tougher times.

Audiomack playlists blasting, I've spent much time since 2020 cruising around the island, with my friend Alexesie, a 27-year-old photographer steeped in current island music and always searching for new sounds. I was surprised at the noisy vacuum left in these hot new tunes by the absence of what we lovers of "one drop" reggae treasured most—the bass. More unsettling still, this bass-less genre seemed to dwell on the "bad mind," malicious backstabbers who smile in your face while plotting your downfall; it was more nihilistic than emo. While Jamaican music has evolved in countless ways since the '70s, the shift from that era's conscious "roots" sound to the faster, rougher dancehall styles of the 1980s and '90s didn't involve quieting the music's low end. In today's dancehall records, by contrast, the bass is a ghost of its old self, suggested by a low drone or irregular pulse, if at all.

But as it often goes with new sounds, after a while I started to adjust to and even enjoy the thinner, tinnier music, despite its excessive use of vocoder. I was struck by three artists in particular. Rygin King lured me with plaintive tracks like "Song Cry," and "Broken," sung from the wheelchair he's used since he was paralyzed by gunfire at a show. Then there was Shane-O. His anthems like "Dark Room," "Wicked People," and "Hold On," tuneful, heart-stabbing evocations

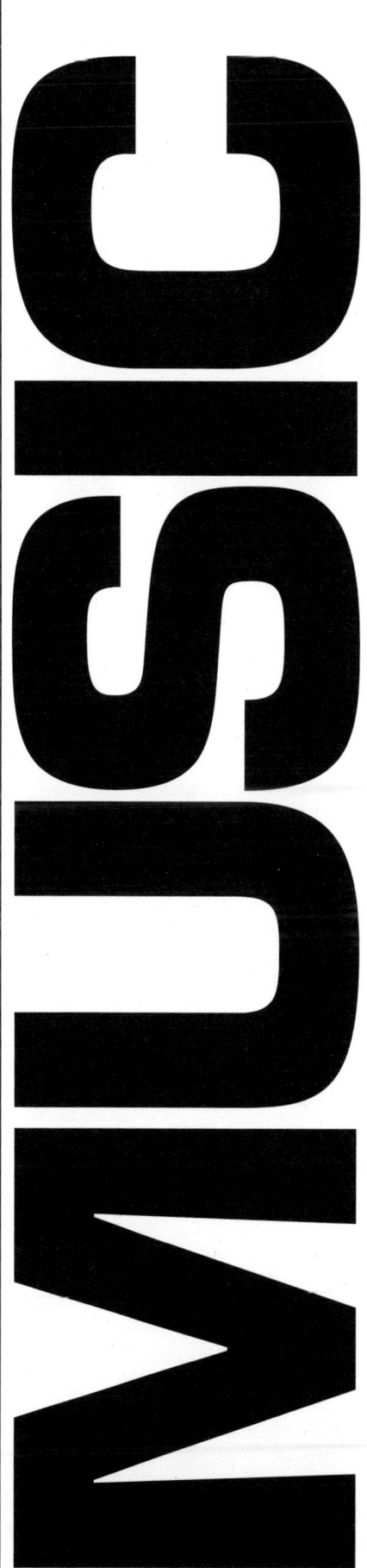

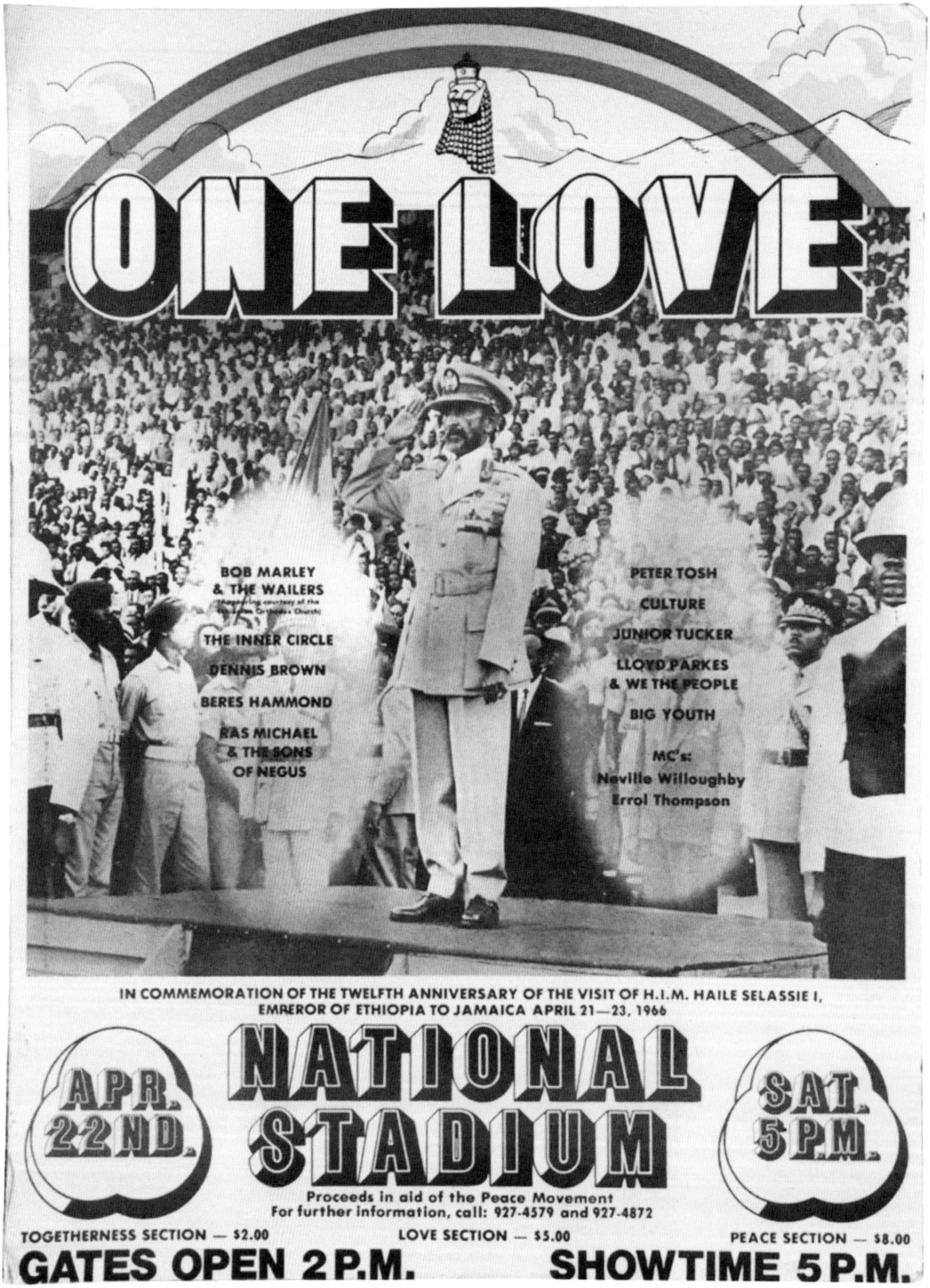

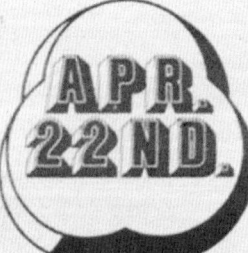

ONE·LOVE

BOB MARLEY
& THE WAILERS
(Appearing courtesy of the
Tuff Gong Organization)

THE INNER CIRCLE

DENNIS BROWN

BERES HAMMOND

RAS MICHAEL
& THE SONS
OF NEGUS

PETER TOSH

CULTURE

JUNIOR TUCKER

LLOYD PARKES
& WE THE PEOPLE

BIG YOUTH

MC's:
Neville Willoughby
Errol Thompson

IN COMMEMORATION OF THE TWELFTH ANNIVERSARY OF THE VISIT OF H.I.M. HAILE SELASSIE I,
EMPEROR OF ETHIOPIA TO JAMAICA APRIL 21—23, 1966

NATIONAL STADIUM

APR. 22ND.

SAT. 5 P.M.

Proceeds in aid of the Peace Movement
For further information, call: 927-4579 and 927-4872

TOGETHERNESS SECTION — $2.00   LOVE SECTION — $5.00   PEACE SECTION — $8.00

GATES OPEN 2 P.M.   SHOWTIME 5 P.M.

Original One Love concert poster, 1978. Art Printery LTD, Kingston, JA.

of loneliness and angst, namecheck "bad mind boys," yet suggest, with surges of female harmonies, how they might survive seemingly impossible obstacles and live to hit a higher note.

Chief in this new crew was Rebel Sixx, aka Kyle George—a prime purveyor of the new "Trinibad" dancehall sound, who I learned was not Jamaican but Trinidadian: a Trini from the southern Caribbean island known as the wellspring not of reggae but cheerful, friendly-sounding calypso and soca. Rebel's cynical poignancy on "No Trust No Love," sung with Travis World, reeled me in. It was a breath-stopping shock when I heard that he was already dead at 26; his ironic soprano was reaching me from the grave. I was determined to learn more about his fate.

* * *

My need to find out more about Rebel Sixx's story was propelled, no doubt, by events I lived five decades ago, and that I later chronicled in my *Book of Exodus*—an account of the attempt on Marley's life, his escape to London where he made *Exodus*, and his return to Jamaica 18 months later for a historic peace concert.

The idea for the concert was hatched by two rival dons—as the gangster-bosses of Jamaica's "garrisons," or impoverished neighborhoods shaped by political patronage, are known here—who found themselves in the same jail cell. Claudie Massop and Bucky Marshall connected the dots to see how, as neighbors and enemies, they had been manipulated into trying to kill each other. Both "bredren" of Marley's, they called on Bob to risk returning to the island and playing a "One Love Concert" for peace—and he paid for their tickets to come to London to discuss it.

In the mid-1970s, the loyalties of Jamaica's two-party system were clearly aligned with Cold War interests. The Peoples National Party (PNP), headed by the glamorous Michael Manley, was involved with Castro's Cuba and Russia, whereas the Jamaican Labour Party (JLP) headed by Edward Seaga (also known as a record producer and student of African rituals) was tucked in bed with America. Kingston neighborhoods like Marley's Trenchtown (a PNP garrison) and his childhood friend Claudie Massop's Tivoli Gardens (a JLP stronghold) were warring no-go zones whose residents could not cross into neighboring streets. The strings of the puppets performing gang warfare in the rundown ghetto were being pulled from far away, but the bullets and killings were real.

So the dons' initiative was bold: an attempt to throw off the political shackles that were holding Jamaica's poor captive. The exhilaration of the Peace Concert, and the Rasta drums that throbbed through the city all night, will always stay with me. The music was magnificent and a joy for bass fans. Robbie Shakespeare brandished his bass like a spear as he backed Peter Tosh, and the inimitably rounded notes of Family Man's instrument anchored Bob's Wailers. Bob called the rival political leaders to the stage and made Manley and Seaga clasp hands over his head. Mystically, and as if by sympathetic magic, lightning flashed, thunder rolled, and the moon shone red above their heads. The truce may have only been theater, but the image was indelible.

Deep in the throng pressing against the stage, I was entranced and permanently infused with the idea that music can bring about change—even as I fought off someone trying to steal my cassette player. The Peace Concert of 1978 has stood since as a beacon of possibility for those who believe that artists have a role in bringing change to their community and the world.

In the weeks after the concert, Bob's friend Claudie Massop was just one of the peacemakers killed by the police. There was a widely-held rumor, often now mentioned as fact, that one of the political parties smuggled guns inside the sound equipment brought in from the States for the show. The undercurrent of violence simmered beneath the release of peace, even then.

Nowadays in Jamaica, the PNP-JLP rivalry persists. The murder rate is still formidable. Around the world, as in Jamaica and the wealthier island of Trinidad, the sway of powerful gangs has grown in recent decades, with all the gangsterism that implies. With COVID and its repercussions intensifying the mix, all these elements shake up into today's toxic "Bad Mind" cocktail, arguably even more lethal than when Bob got shot.

2.
BAD MIND

Must they always kill the best and brightest, or try to? Rebel Sixx foretold his demise in music, like Tupac and Biggie. But his murder, for me, reverberated with Marley's narrow escape and the death of the Wailers-style bass—killed by changing tastes and technology. Somehow the bass came to represent not just a sonorous absence, but an absence within what I've come to think of as the Bad Mind movement—that sense of militant, against-all-odds, revolutionary optimism, which few of this popular generation of artists seem to feel. In the old conscious reggae crew, Bob Marley was not the only one to leaven his most dire critiques of humanity with enough hope to keep you going. Bad Mind music is the offspring of a generation several times removed from Bob's. They were raised on digital music amid the drip of trickle-down financial policies and the chasm of the economic gap.

Before the island won independence in 1962, Jamaica's music was the mellow, folk-y lilt of mento, which sped up to the urgent scurry of ska in the 1960s, inspiring Britain's lauded 2-Tone movement—The Specials, Selecter—and a worldwide community that still flourishes today. Live DJs equipped with giant sound systems—"toasting" over twelve-inch disco mix records at Kingston street dances—are generally acknowledged as being forerunners of American hip-hop. Conscious reggae was spearheaded by the leonine Bob Marley, while the masters of dub, its ghostly sister, turned the studio into a tool for crafting records from a post-modern collaging of echoes and beats. Both were grounded in classic organic rhythm sections of bass and drums, until the rise of synthesizers and other digital tools eliminated the need, for many music makers, to play live instruments at all.

The big shift came with Wayne Smith's "Sleng Teng" in 1985, an iconic hit whose reedy, sugar-high Casio loop was designed by a Japanese woman, Okuda Hiroko, and tweaked for reggaematic purposes by the producer Prince Jammy. After that, the deluge of synthesizers, along with an increase in guns and cocaine on the island, saw the music become ever more crude, urgent, and violent. Dancehall went on to be the impetus for the Latinx reggaeton sound. Born in Puerto Rico and elsewhere of Jamaican parents, reggaeton dominated many of the world's dancefloors until Afrobeats emerged from the motherland, Nigeria specifically, a few years back. All these post-digital sounds share the bass-lessness of Jamaica's Bad Mind hits.

Haunting and haunted, these songs sound like premature requiems, *de profundis*, coming at us live from the depths. Was hope alive? I hoped so. Hope had to be there, somewhere, as ultimately, humans cannot live without it.

* * *

The producer most responsible for the Bad Mind sound, I'd heard, hailed not from Kingston but rather a volatile area outside Montego Bay called Salt Spring, on the other end of the island. Born Andre Whitaker, he was better known as Squash. Regarded as charismatic and creative, he posts images to TikTok that highlight his puppy dog eyes, peering from his heavily bleached, tattooed face—the in look for "bad boys" in Jamaica—and he poses endearingly with his young children. He also ran a notorious gang, the 6ix, who were as well known for violence as he was for toothsome beats. That was as much as I had to go on.

But digging into the deeper matrix of the Bad Mind sound would prove to be one of the trickiest music stories I've ever tried to report.

Squash's unavailability was easy to understand, as right then he was in jail in Florida on ICE charges. But almost everyone I tried to speak with was wary, I quickly gleaned, about what the Jamaican media have dubbed the Salt Spring Gang Wars: a steady barrage of killings, over recent years, in this and other garrisons also known for music. Salt Spring is Squash territory, just as its next

and rival town, Flanker, is known as the stronghold of Tommy "Sparta" Lee. Interviewees displayed a reluctance I had never encountered in six decades of music journalism. They suddenly left the island or stopped returning calls. People didn't want their names used. Was I living one of my beloved Agatha Christie mysteries?

First to vanish was Rygin King, whose melodic track, "Song Cry," always reached out and touched from the car radio. Rygin had recorded with foreign rock stars like drummer Zak Starkey and his singer wife, Sshh Liguzz, who put us in touch. And we did speak briefly. Rygin told me of his busy schedule in America, where he's been living near Miami, still doing concerts while undergoing physical therapy. He confided his interest in writing a book about his experiences; he seemed eager for any help I could give. It seemed worthwhile. He was determined to use his fate as a positive example. But when time came for the interview, Rygin ghosted me—an unusual move for an artist on the make, and one that made me paranoid *he* was feeling paranoid.

Luckily, a friend of a friend introduced me to a source close to the various communities and investigations key to this story. This crucial guide, who hereinafter shall be known as Nameless here, explained to me that the 6ix crew, like most Jamaican gangs, generally liked to keep their wars confined to their own HQ; the Jamaican police and Defense Force are in a valiant struggle to make sure it stays that way, and that the fighting in Salt Spring and nearby doesn't spill over into the prosperous tourist hub of Montego Bay—even as the 6ix crew has succeeded in spreading its gangsterism to Atlanta and Miami.

The 6ix was a formidable two-headed creature, it seemed. The gang was also a sound, and it was Squash's "riddims" that had inspired Rebel Sixx. His belief in the 6ix seemed absolute; most of his songs start with a sibilant utterance of the numeral. But before I tried to make my way to Salt Spring—with the guidance of Nameless and Alex at the wheel—I sought out Shane-O in Kingston. Buoyed by banks of harmonies, I had already begun to turn to his music when fending off despair.

* * *

Bad Mind ghetto dramas were lived material for both Rebel and Shane-O. Rebel's mordant irony compelled me, while Shane-O's swelling melodies moved me differently. And Shane-O was accessible. Having grown up in a garrison called Common Pen, at the foot of a steeply climbing hill in Kingston, his success has seen him move to its breezy heights.

In the 1960s, "garrison" emerged as Jamaica's territorial term for communities built by one or other of its rival political parties, to house residents of slums cleared to construct them. These neo-Stalinist concrete enclaves, replete with violence and absent licit economic sustenance, became fiefdoms controlled by gangsters who

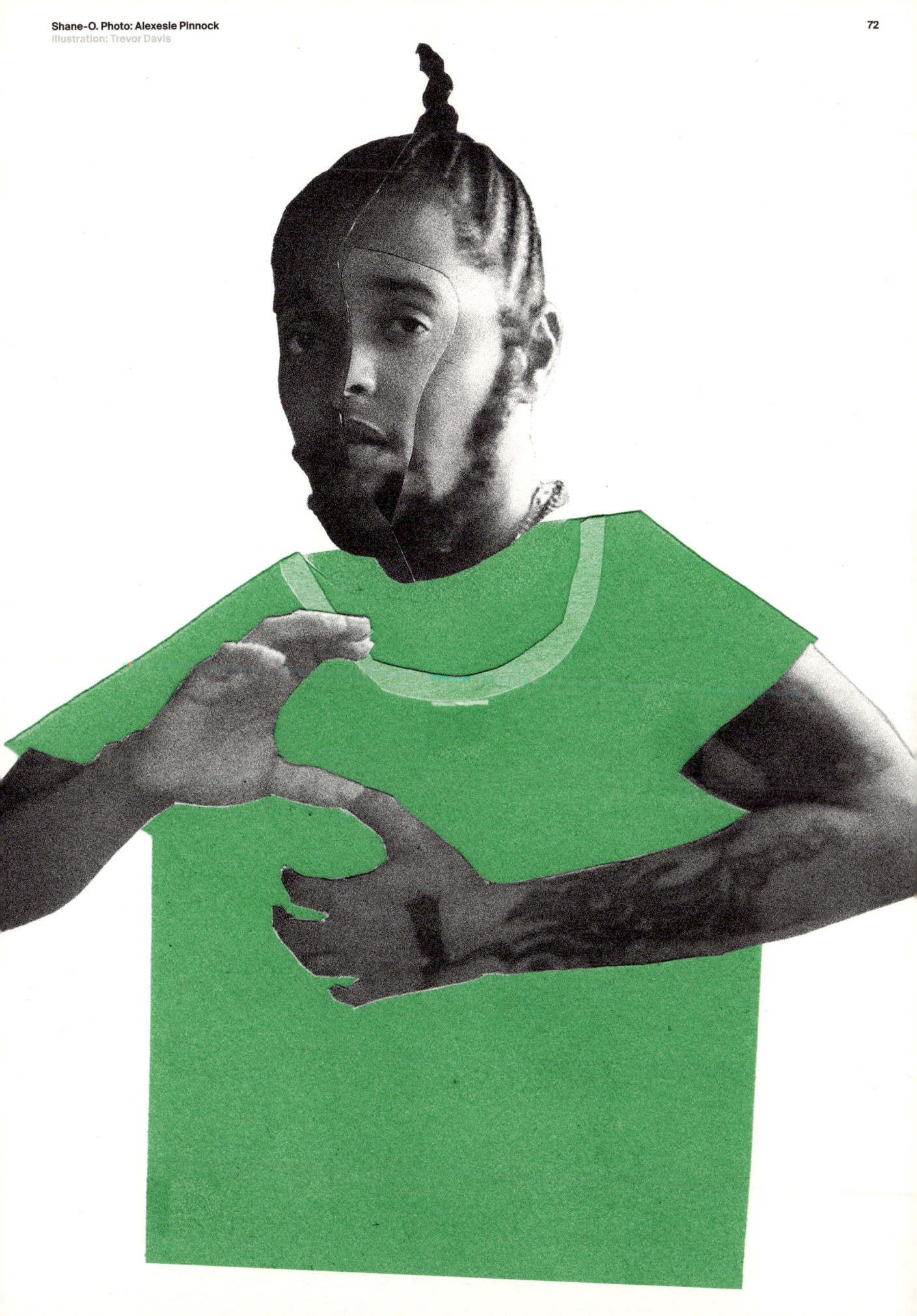

do their political patrons' dirty work, leaving the manicured hands of those who run the system clean. The boat is only rocked when the dons start to think twice about who it is they report to, or look for better deals elsewhere, on or off the island. The notorious Kingston don Christopher "Dudus" Coke, whose father was implicated in the attack on Bob Marley, ran the old JLP garrison of Tivoli Gardens until 2010. His extradition to an American prison made international news and embarrassed Jamaican authorities; not only did Coke and his Shower Posse enjoy the support of the country's people, but it was revealed that Coke had links, through legitimate

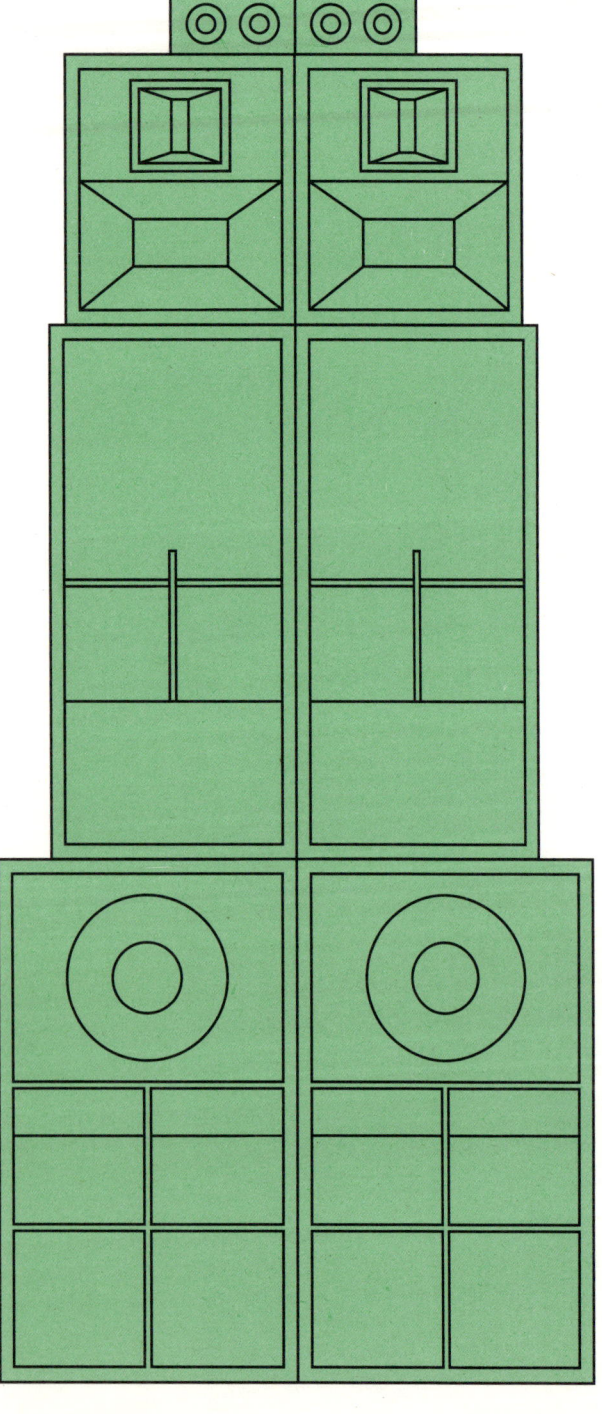

businesses he owned, to the Jamaican government. As much as a garrison can be an armed fortress, it can also be a culture, community, and home, as Shane-O told me of his own, Common Pen.

An engaging character with shrewd yet merry eyes, the singer of "Hold On" has a jaunty uni-plait hairdo and an affinity for cartoon graphics t-shirts. Shane-O's lighthearted affect, as we spoke, countered the emotional intensity of his videos, in which he's seen listening carefully to his grandmother's advice; wailing his Bad Mind choruses, a solitary anti-hero ranting on a pier against a turbulent sea; or dropping to his knees, frenzied, in the middle of a busy highway.

For Shane-O, those skills were absorbed early on. Raised in poverty by his mother, who often didn't have food or water for her kids, he credits his mostly-absent father—a skilled singer and DJ in Common Pen—for his musical talent. His lifeline was an older youth called Sno-Cone, who would help him out if he had nowhere to eat or sleep, and took him in the studio to record his first single, "Rice and Peas," aged 11. Big artists like Beenie Man and Bounty Killer were there, but Shane-O wasn't nervous. "Mi haffi do what me haffi do." The record did well and he went on to perform, that same year, at Sting, an annual showcase for dancehall's latest and greatest.

In this villa that he rarely leaves except for shows, he plans to install a recording studio. It feels like a communal "ranch," with sundry dread bredren making it their HQ. He coos over the white pigeons who take pride of place in his backyard. And then he shows me to his ample villa's cave-like basement. It's a utilitarian room I recognize from his video for "Dark Room," a fraught, luscious song that's among my favorites.

As Alex and I listen to Shane-O perform "Hold On," then "Distress of Mind," acapella, he sounds out the chorus, then explains, "When the lights turn out, it is murder, official, a whole heap of bloodshed. That's the place where we ghetto youth grow up. I always sing 'dem kind of music, to hold on tough to life, as long as life lasts."

Why do so many of his songs revolve around Bad Mind? Shane-O snorts with appreciative laughter at the analysis, then gets serious. "It's people [who] try to fight and hold down other people, when they don't even know a person. Nuff people me' a meet, and me no' know what a gwan, but it becomes a strong fight, bad mind—active! When you' talented, it terrible."

I ask if he knows much about Rebel Sixx. No luck, though he appreciates his music and knows of his fate. He shakes his head. "I live like this, often in a room by myself," he says. "So me just live, because on earth, you have to work with it while it goes by."

It felt like he was echoing Marley telling me in 1976, "Me is not a man who move up and down too much." Days later, gunmen had assailed his house with fatal intent. As we're leaving Shane-O's yard, I wonder if there might be another Peace Concert, like Marley had? "It no gonna work," he replied. "The government alone can do it, but they themselves have their own wickedness them must deal with." But hope is a must, Shane-O insists. "If it used to be nice before, it shall be nice again. The bass and the one drop soon come back," he says. "So it go."

Heartily encouraged by our session, I mention to Shane-O, before his wrought-iron gates close behind us, that Alexesie and I plan to visit Salt Spring. His face grows somber. "Salt Spring is rough," he warns. "Remember, dem have a curfew."

3.
SALT SPRING TO
ST. JAMES,
DANCEHALL TO TRINIBAD

Survival is serious business for prominent artists like Shane-O. On a recent episode of a popular online show devoted to Caribbean music, *The Fix*, the host asked singer Jahlanno, a friend of Rebel Sixx, how he "maintains neutrality." Dapper and astute, Jahlanno had his reply ready: "I don't try to be the best. It comes with a lot of things. Once you get to the top, you still have to keep climbing and don't fall. I just try to do my thing, the way I do it."

What a signal to send for survival. But also canny advice, it was clear, on this island where the epicenter of the bass-less Bad Mind sound appeared to be one of its most notorious garrisons: Squash's home area of Salt Spring, a few hours' drive from Kingston and reachable via a turn-off from the highway that lines the island's beautiful north coast, leading to Montego Bay.

The drive along the North Coast in the parish of St. James curves through a series of mellow bends beside the turquoise ocean, or what you can see of it between high mansion walls. This is the Caribbean's true Gold Coast, a series of towns whose Spanish or English names—Ocho Rios, Gibraltar, Salem, Rio Grande—remind us of all the blood shed here in the serial conquests of invaders from Europe. The Taino or Arawak Indians, Jamaica's first nations, were genocided within fifty years of the Spanish arrival, followed by 400 years of subjugation under varying levels of sadistic brutality or social suppression under the British. By the side of the road, old water wheels that kept the plantations turning are now historical landmarks. Fantastic examples of workmanship and engineering, they are a testament to the skills of enslaved people, and still so valued that they have not—as one might think—been torn down stone by stone.

After independence, and a collapse of the global market for the one natural resource—bauxite—in which Jamaica is rich, the island's leaders decreed that tourism

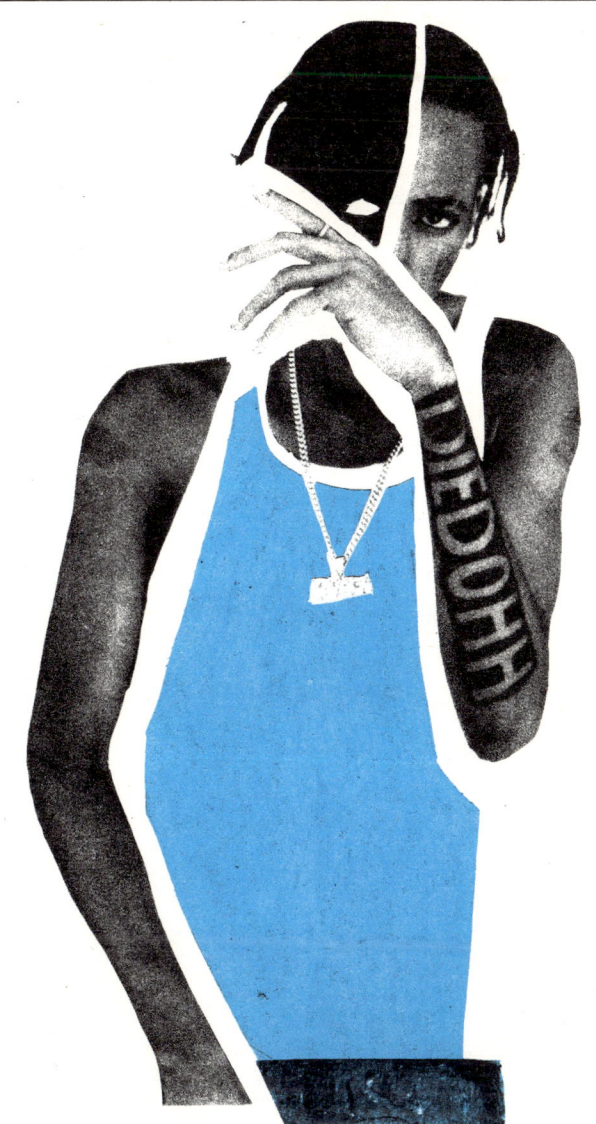

Diedohh in Flanker, Montego Bay, Jamaica, 2023. Photo: Alexesie Pinnock
Illustration: Trevor Davis

was to be the key to financial independence. Shifts in the weather and the world economy alike can torpedo income from tourism. Within the hotel industry that dominates the North Coast, there are opportunities for jobs, but often no pay for workers in the off-season. There are energetic Jamaican-owned resorts like Sandals and Couples, and characterful boutique hotels like GoldenEye and Jake's, but most of the money earned by big resorts owned by American or Spanish conglomerates leaves the island. The stubborn colonial mentality that shapes much of the tourist trade is also reflected in the fact that Jamaica's residents don't have access to many of their own beaches, carved up as they are between billionaires, foreign governments, and hospitality chains. And an island dependent on tourism as its main "industry" is vulnerable, like an aging courtesan dependent on fading charms.

But I'm reminded as we drive the North Coast that at least Jamaica's beauty is eternally ravishing. Tourists are needed, and there is plenty of reason for them to come;

the ghetto incidents depicted here won't touch them. In Ocho Rios, we pass a mural of the North Coast's musical godfather in the 1970s, Jack Ruby, bumping fists with his bred'ren, Bob Marley, born nearby in the "garden parish" of St. Ann. As we pass towering cruise ships, Dunn's River Falls, the Dolphinarium, and enter St. James, nearing Montego Bay, I recall what Nameless said about the workings of the underworld in this parish. Known as an epicenter for credit card fraud and other financial scams, the parish's local Jamaican perpetrators, targeting marks around the world, have become infamous. "St. James's persons whose wealth comes from areas other than music," he told me, "use music to legitimize someone with potential. Scammers and people who do fraud will back this person as their producer or manager, laundering their money. Almost all the gangs in St. James are connected to a musical person."

Salt Spring, I understand, is a JLP area run by Squash. The nearby garrison of Flanker, Nameless has told me, is affiliated with the rival PNP. Both areas have recently been ravaged by murders aggravated by splits among the gang factions. Gangs don't only have friction with those they attempt to dominate; they have intra-cell beefs that cause them, like amoebas, to continuously divide.

Gang-associated musicians, whichever team they bat for, often release celebratory tunes describing local murders with grim detail only insiders would know. Gangs lure young artists by giving them access to popular "riddims" to sing over—with the ever-lurking danger they'll get in so deep they can't back out. While St. James has always had an underground economy. "The trade kept on re-defining itself," Nameless told me, "Before it moved to cocaine, back in the day when marijuana was illegal, the Rose Hall stretch of straight road would be closed for 5-10 minutes, a plane would land, load up with ganja, take off, and the traffic would move again."

His words echoed in my ears as we drove down that very same scenic strip, past the lush green lawns of the Rose Hall Golf Course, some miles outside of Montego Bay proper—and then Alexesie's phone rang. It was the Old School Friend. Something had come up. Despite our three-hour drive, he couldn't meet us after all.

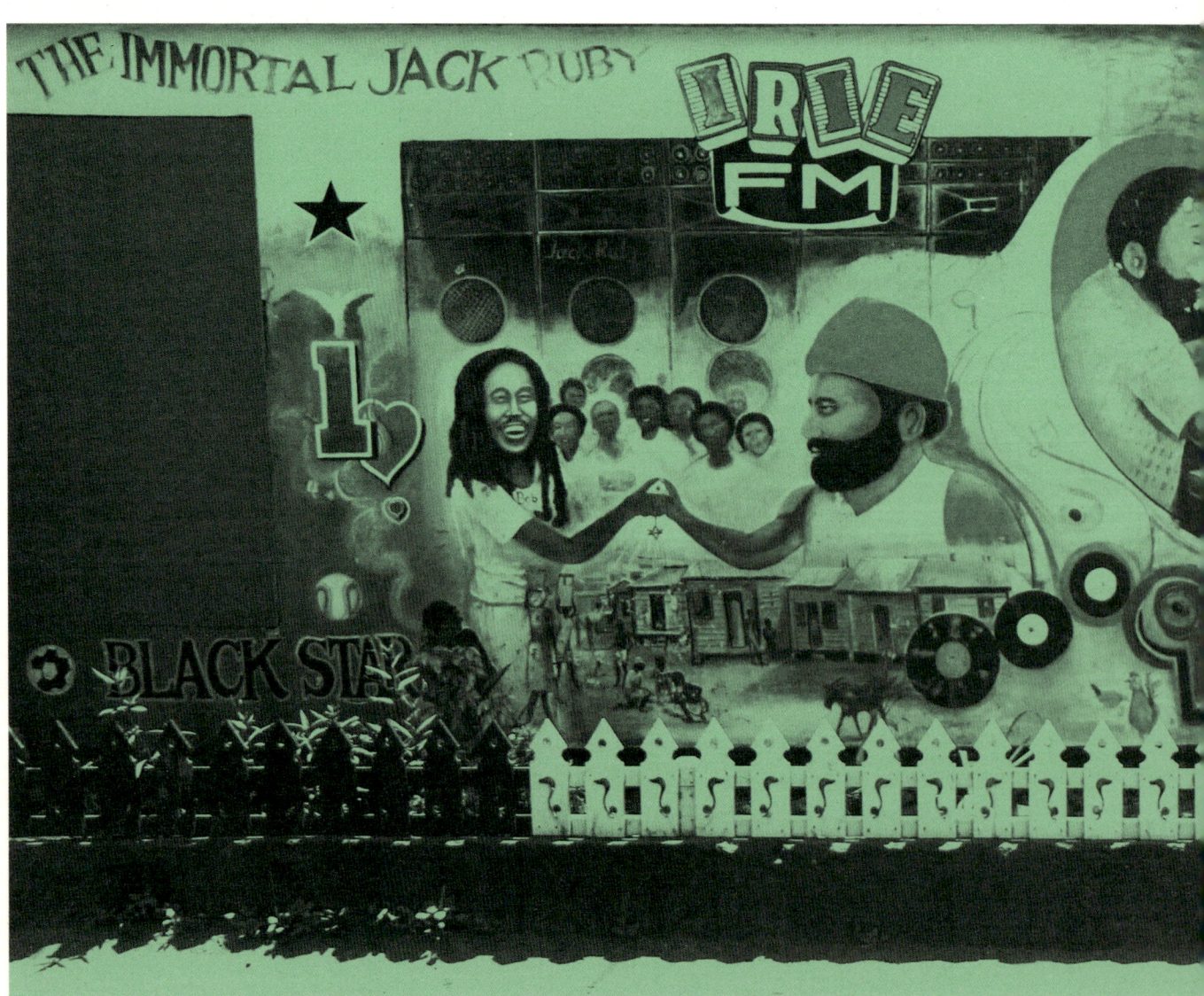

Jack Ruby Ave, Ocho Rios, Jamaica, 2023. Photo: Alexesie Pinnock

Nevertheless, we had to try and get at least some sense of what Salt Spring looked and felt like. Its energy propelled the dance sounds that turned so many of its youth into "soldiers" of various kinds—some combat-ready—but many conscripted into armies they never wanted to join. Three miles before the turn-off for Salt Spring, we took the road heading to Flankers—its rival garrison and the home of Tommy "Sparta" Lee—instead.

* * *

Salt Spring, like much of coastal Jamaica, was a plantation in slavery days when the phrase Bad Mind was born of the horrors its residents endured. In 1774, the sugar plantation at Salt Spring belonged to one Joseph Bowen, and produced 93 hogsheads of sugar; it was the enforced home of 151 enslaved individuals as well as four men bearing arms and eight "women or children." The human trade, of course, is the Great Crime, bigger than scamming—the primal scar that, along with the death of the Tainos, wounds Jamaica and with it, the world. It's perhaps only fitting that this island, once known as the

"Reaper's Garden" for its brutality, is now also where Tommy "Sparta" Lee invented what's known as Gothic Dancehall.

Since his recent release from a two-year stint in prison on gun charges, I learn Lee has been laying low; even his own co-workers can't track him down. With his music's allusions to horror films and devils, the leader of the "Sparta Army" is a divisive figure in Jamaica, despite his huge following. On this island full of churches, he invited accusations of Satanism by releasing the song "Daddy Demon," and adopting the title as his alias. Then there's his track, "Psycho," familiar to players of *Grand Theft Auto 5*, for whose video he donned a slasher flick mask. It seemed unlikely that we would get to talk to Tommy Lee, but some miracle might make it happen.

We drove up the road toward Flankers until we encountered signs of life; a bar with a tree-filled, shady garden on the right and what looked like a lumber yard on the left. Behind this comparatively prosperous outskirt, I knew, there sat a shanty town—fetid, oppressive tenement alleyways and tumbledown shacks strapped

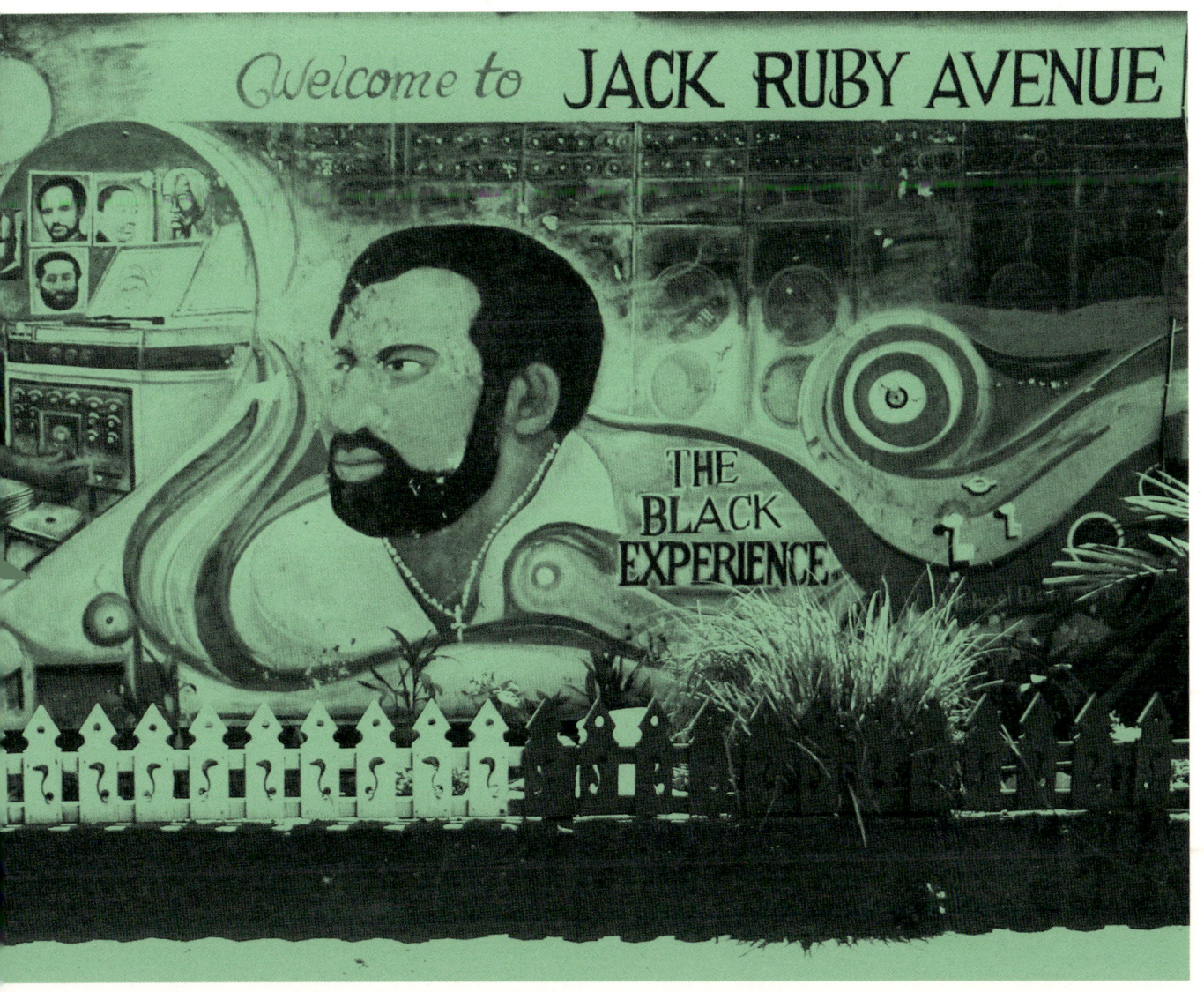

together with flotsam of the garbage heap, zinc if you're lucky. But around us, Flanker seemed to be booming.

As if determined to contradict the area's bad rep, locals in the inviting yard of the small but colorful bar we strolled into gave me a major welcome. I was invited to sit, play dominoes, have a beer. Then I mentioned our mission to visit Salt Spring. Oh, dear. Heads were shaking and frowns all round. "You'll never get in there," said a domino player, just as Alex called me over from across the street. By the lumber yard there, a group of guys included a friendly man in neat jeans who insisted: "Forget going to Salt Spring. You won't be allowed in."

Our new friend, it turned out, was a cop who informed us we'd just missed Lee himself. The compact car we'd seen pull out of the lumber yard just as we'd entered contained the man we sought. Perhaps he'd been warned, by some angels or devils, of the media's impending approach, though we hadn't known it ourselves.

Tommy Lee's Gothic Dancehall is designed to send chills that are both a reflection of, and a diversion from, IRL violence. For good or ill, Lee's nephew Diedohh, who happened to be around, demystified his uncle's

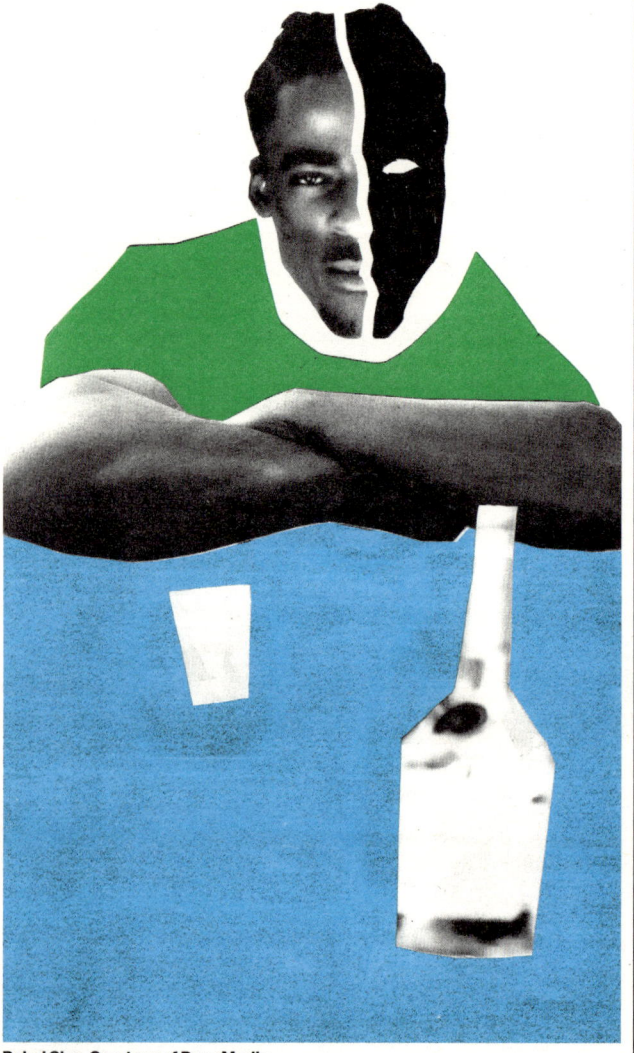

Rebel Sixx. Courtesy of Raya Media
Illiustration: Trevor Davis

demonic presence. "It's an image he created, not that personal." Despite all the hissing and scary gesturing, Tommy Lee seems to prefer a quiet life. He had changed his name to Sparta, and built his Sparta Army, as a declaration of independence from former associates like Vybz Kartel—the iconic dancehall artist with a genuinely demonic image, who landed in prison after a charred body was found in his yard. Vybz, who was recently released from jail, hails from a section of the Kingston-adjacent city of Portmore that's known as Gaza City.

I mentioned the coincidence of the original Gaza Strip, now devastated by war, being in the news; Gaza's image of beleaguered volatility is what inspired Jamaican ghetto-dwellers to adopt the name. Diedohh hasn't heard about it. "As Jamaicans, we are locked off from knowledge of those things, though I have a little taste." As a professor, I suggest he look it up online. One North Coast neighbor reckoned it was natural that a far-off war shouldn't engage his interest when Jamaica, and this corner of it, is so riddled with domestic ones. But Diedohh is also insistent that, despite the regular shootings and his grandmother's warnings not to play with kids from Salt Spring, Flanker and its neighboring garrison aren't at war. "Every community have a little bad side to it," he says. "Flanker and Salt Spring [people] both have bad intentions. We have bullies in common, but we also have good people." What most perturbs him these days is the public curfew that's been instituted in both Salt Spring and Flanker, to try and curtail shoot-outs at night.

"It affects us as young people," he explains, frustrated. "And it really drains our energy as artists. Often when we reach the show, we find police have locked it off." But this 21-year old, proud that his music is streamed in both Flanker and Salt Spring—and, he says, "in all 14 parishes"—is also hopeful that music can bring peace. "Music is life and music can make a difference," he says. "I hope that my music is a message of positivity." He also hopes, he says, to one day work with Squash in Salt Spring. "He's a friend of my uncle."

This is surprising, but perhaps it shouldn't be. Hadn't deadly political enemies Claudie Massop and Bucky Marshall brokered Marley's Peace Concert back in the day? With island gossip and the music industry both invested in their rivalries, it may be too complex for Squash and Sparta—even if they privately enjoy each other's company—to officially hang out.

Nameless has explained to me that Squash's own backstory and involvement in crime were both shaped by his elder brother—a kingpin of scamming here, who was killed by police before Squash took over his 6ix gang. A divisive and complex figure, Squash has been cleared of scamming charges here, but tainted by incidents like a recent double murder committed by a producer from his crew that was caught on video. He has told journalists that police harassment drove him to move to Florida for a while; others claim the attacks were from his reputation as a "violence producer" of "murder music" and the scammers who seem to surround him as they did his late big brother. Either way, with Squash unreachable in prison I turned to one of his associates, the producer Dindin of Hemton Music, to try and untangle the story of what befell the great Rebel Sixx.

"But it really isn't me you should ask," he insisted. If I really wanted to understand Rebel Sixx, Dindin told me, I needed to direct my queries to another island.

* * *

Trinidad and Jamaica are both Caribbean islands colonized by the British, but in key ways that's where the resemblance ends. Jamaica is bigger; Trinidad richer. Jamaica is mostly Black; Trinidad's population pie is split three pretty equal ways: Black, Indian, and Everybody Else. Trinidad's economy, like that of its sibling island Tobago (the nation-state of Trinidad and Tobago is known as "T&T") was long shaped by the unpaid labor of captive Africans, and also by indentured workers from India. But T&T's modern fate has been carved by geology. Once attached to Venezuela—and still attached, in terms of being an entrepot for its cocaine and guns and human trafficking—Trinidad has oil resources that mean it has long been awash with petro-dollars, depending on the market. Trinis, on average, earn about three times as much as Jamaicans.

Musically, satire has dominated sequential genres of witty calypso, happy, hearty soca, and Trini rap, known as rapso. Starting in the 1940s, Trini calypso was so big worldwide that people thought Lord Invader's "Rum and Coca Cola" was actually by the crystal-voiced blonde Andrews Sisters—though their international cover hit was rarely understood as referring to local sex workers on the US Navy base built there during the war.

The famous steel pan chimes the sound of rebellion. In this place where drums were once banned by the British, the "pan" was forged in the 1940s from oil cans in the still turbulent Laventille area of Port of Spain. In the 1980s, soca star David Rudder scored a conscious hit with "The Hammer." Since then, Trinidad has become a much more violent place—along with everywhere else. In pace with "urban" American music, in the 1990s Trinidad's response to hip-hop was rapso; today's exponents of the "Trinibad" sound project themselves as musical gangsters reflecting Jamaican dancehall through the prism of Trini creativity.

Above all, on this island that's been shaped by Catholic culture since colonial days, the sexy glitter of carnival—marking a "farewell to the flesh" before the onset of Lent—is the cultural hub. Climaxes of carnival, like the Soca Monarch Competition, consume the island and magnetize international Trinis to do whatever it takes to get home. Carnival has also long been the main focus and inspiration for Trinidad's music. But in recent years, as crime and anxiety have come to dominate the island's culture, that's changed. On this island where "liming," an active form of chilling, is a national pastime, party and events promotion are key activities for many gangs. So are sex trafficking, money-laundering, and sand-blasting for concrete used for endlessly ongoing construction projects, many of them born of crime that's organized, disorganized, or governmental.

By contrast, in Jamaica, and especially on its north coast where Squash is based, the beleaguered economy is based on tourism and the scamming which, paying for much of the country's musical output, has itself become an art form requiring ingenuity, teamwork, guile, and expert acting skills. In fatter Trinidad, many prominent gangsters and gangs double as "community leaders" who kill to secure government contracts in areas like St. James, the part of Port of Spain where Rebel Sixx grew up.

A wry intelligence in his voice first drew me in. The way Rebel Sixx sang it, bloody images became a sad seduction, leading to contemplation; his sweet and sour

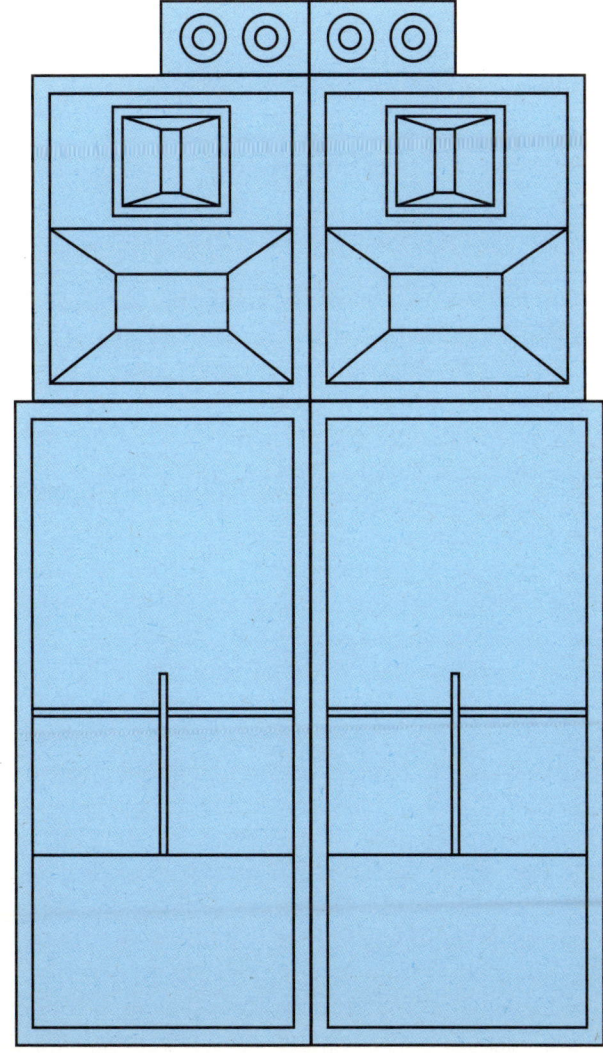

sound poetically, cynically, recounting brutality with a honeyed soprano and heavenly timing.

He was a musical original, and I had no idea when I first heard his music that he was dead, let alone the circumstances. When I fell for Rebel like any fangirl, I assumed he was Jamaican as he was so popular there. During that year I spent back on the island, his songs were everywhere—the bleak delicacy of his and Travis World's "No Trust No Love," and "Rifle War," "Ghetto Prophet," "Message to the Heart." As I adjusted to the feel of this music, oddly bass-free and with an alarming emphasis on the evils of Bad Mind, realizing my favorites were by one artist helped me to absorb and understand the new sound.

Laconically, he loped through his videos of bullets and betrayal with self-effacing detachment, as if to say: look at the craziness we have let ourselves get into. Along with the music, it was the philosophical sorrow in Rebel's eyes that compelled me to explore his story. Whether in interviews or music videos, even when surrounded by a dancing, drinking crowd, his eyes signaled that he was running—making music—as fast as he could, trying to keep ahead of the doom he could sense was chasing him down.

4.
THE GHETTO PROPHET

To understand the hellhounds on Rebel's trail, and his songs' roots, Dindin told me, I had to talk to his people. "Find the one who knew Rebel best," he said. "His producer in Port of Spain: El Faltino."

Seen over Zoom sitting in the shade of a white-walled building, El Faltino, aka Asim El-Salih Faltine, was a humorous, quick-witted man with a boho uni-dread. He'd met Rebel in 2019, he told me, on the Friday after Carnival. "We were on a musical journey together." El Faltino's area, Belmont, is the birthplace of Stokely Carmichael, the 1960s Black Power activist. But the foremost artist it had recently produced when Rebel came on the scene was Faltino's artist K-Lion. Born Kwinton Thomas, the sparky young man's light, skipping vocals were pushing a new strand of Trinibad that he and Faltino were calling Zess. Rebel and K-Lion became friends, and went on to record together. "K-Lion was so young, but very talented, one of the few artists we all looked up to. And Rebel was the Ghetto Prophet he sang of, speaking his mind in his humorous way." El Faltino paused. "But a lot was going on. He may have offended certain people…"

The producer's worldview, if not cynical, is sophisticated. "In Trinidad each and every one of us have bad

friends, even if we not bad," he explains, before suggesting that I talk with Rebel's last manager, Hugh Callender.

"Rebel was on a pilgrim's journey," says Hugh, a garrulous, eloquent man whose bushy beard fills my phone's screen when we connect. "And his journey was a big one." Rebel was also, in Hugh's estimation, the foremost exponent of the Trinibad sound.

He tells me about how the bass-less sound of Trinibad, and of much recent dancehall, was born. It's a story, he says, that began in the 2010s, when the heyday of dancehall stars like Vybez Kartel and Popcaan—whose hits didn't want for booming bass lines—began to fade. When it did, around 2016, a Trinidadian producer named Isaac Cozier began experimenting with a sound that joined dancehall beats with soca touches that Cozier laid, crucially, over the bass-less feel of Afrobeats. It was a sound that El-Faltino and others in Trinidad embraced. So did key "riddim"-makers in Jamaica, many of them associated with Squash—among them my contact Dindin and another prolific producer called Automatic who has, Hugh says, spent much time in Trinidad.

The key moment in the transition from Zess to Trinibad, Hugh says, came in 2018. "That's when it went from party music to life music. From light to dark." Artists like Teacher and Swanny were key to this shift. "But no one bus' like Rebel. Like Mount Rushmore, he was the last president—the most influential. With Rebel, Trinibad went from life music to action music. To soldier tales, and stories of war. Stories you only get from living it."

* * *

When Rebel was growing up, Hugh explains, his mother was an ardent church-goer. She made him join her church's choir, where his soprano was prized. But it was in church that he earned the nickname Rebel, because he was the one always pulling pranks. He became a star school athlete and sang in a harmony trio. But his upbringing was not smooth. When Rebel was young, his hard-working mother, a nurse, had to travel a lot and lived for periods in the U.K. His father, a singer called El Negro, was not a stable source of guidance or love, and moved to America while Rebel was very young.

Before he became Sixx, aligning himself with Squash, Rebel was just called Rebel. And as a youth left to raise himself, he created his own family. In his early teenage years, Rebel and his best friend Swanny harmonized together in their harmony group, Z-Tech. Like most youth in their area of Port of Spain, they also became affiliates of a gang called Rasta City. But then Rebel decided to split off. Locals look back nostalgically to that moment just before intra-crew beefs saw Rasta City and their rivals, Muslim City, sub-divide into cliques that sound like numerological codes. Certain "Muslims" became known as "9." After Rebel began identifying with the 6ix, whose members here nurture mysterious ties to Squash and Salt

Tangent

Configuration Space

The space between pegs
connected by twine

representing attacks
that haunt the mind

of the fictional detective
whose colleagues think

he's off on a tangent.

Phase Space

The way a day will phase
quietly

from shock blue
to deep prune

to amber lamp light
and back

in the course
of a few minutes,

maybe seconds.
The units are meaningless.

Pull

"Kinks are sad," she said,

pulling on a thin
string of sensation
like a failed rip cord;
a childhood facial tic.

The disembodied laugh-track
from an old sit-com

as ghost story.

—*Rae Armantrout*

The 6ix were loved and feared. What was their true identity? When it came to the notorious Squash, Rebel was a fanboy. And then he was a prize. In Jamaica, Squash heard and loved Rebel's work, and their online exchange led Squash to visit Trinidad, and Rebel, with members of his entourage.

It was during that visit, Hugh says, that Rebel first turned up to use the recording studio he managed. Cracking jokes but working hard as ever, Squash turned the session into a productive party. He said Rebel had been invited to join him in his capacity as "the only official 6 in the country," as Hugh put it. After all, Rebel had adopted the name Sixx from the music's inspiration alone, before ever meeting the producer who crafted it.

It sounds, I suggest, like being invited to join the Freemasons. Hugh laughs. "It felt almost like that. But you feel very safe with the 6ix, as a creative. Here in Trinidad you have to cross a lot of mountains and rivers to get to where you want to be. But them fellas move with a kind of confidence that makes you feel like it have no mountains. It have no river. It's just flat road."

Rebel's work ethic, on his trek down that road, was also vital. "I have never known a musician to work harder than Rebel," Hugh says. "He would be in the studio at eight and leave at 10, and write two or three new songs a day. He worked even harder than Squash." And the two, Hugh says, remained in close touch.

In a TV interview, Rebel said he spoke with Squash every day. "He is literally like a father to me," he said. "Literally. He is my mentor." From distant Jamaica, Squash became a replacement for El Negro. But in Trinidad, Hugh—businesslike and a big thinker—and El Faltino—the energetic hipster—became Rebel's guides to the music business. "We were trying," Hugh says, "to smooth up his image for the public but still keep it authentic." Under their ministrations, the packaging, marketing, and streaming of Rebel saw "Rifle War" soar to number one on the local Apple charts, and appear on numerous playlists, within four days of its release.

* * *

The more acclaimed Rebel became, the more rumors and negative noises he heard from certain Bad Minds in his vicinity. Encouraged by success, he moved to Bon Air Gardens, Arouca, a slightly "better" neighborhood than St. James, where he had been living with old friends. But trouble followed him.

A cop living across the road had turned his security cameras on his new neighbor full-time, which was more unnerving than reassuring. Then, to Rebel's distress, some of his old Bad Mind acquaintances followed him to the new neighborhood, increasing the artist's paranoia. For protection, he started to surround himself with five or six "soldiers" in the new yard.

Spring, Jamaica, Rebel's old friend Swanny joined with other erstwhile members of Rasta City who started calling themselves "7" and "4"—as if picking the right numbers could be a free pass in the lottery of death. Between Rebel and Swanny, a notorious frenemy-ship was born.

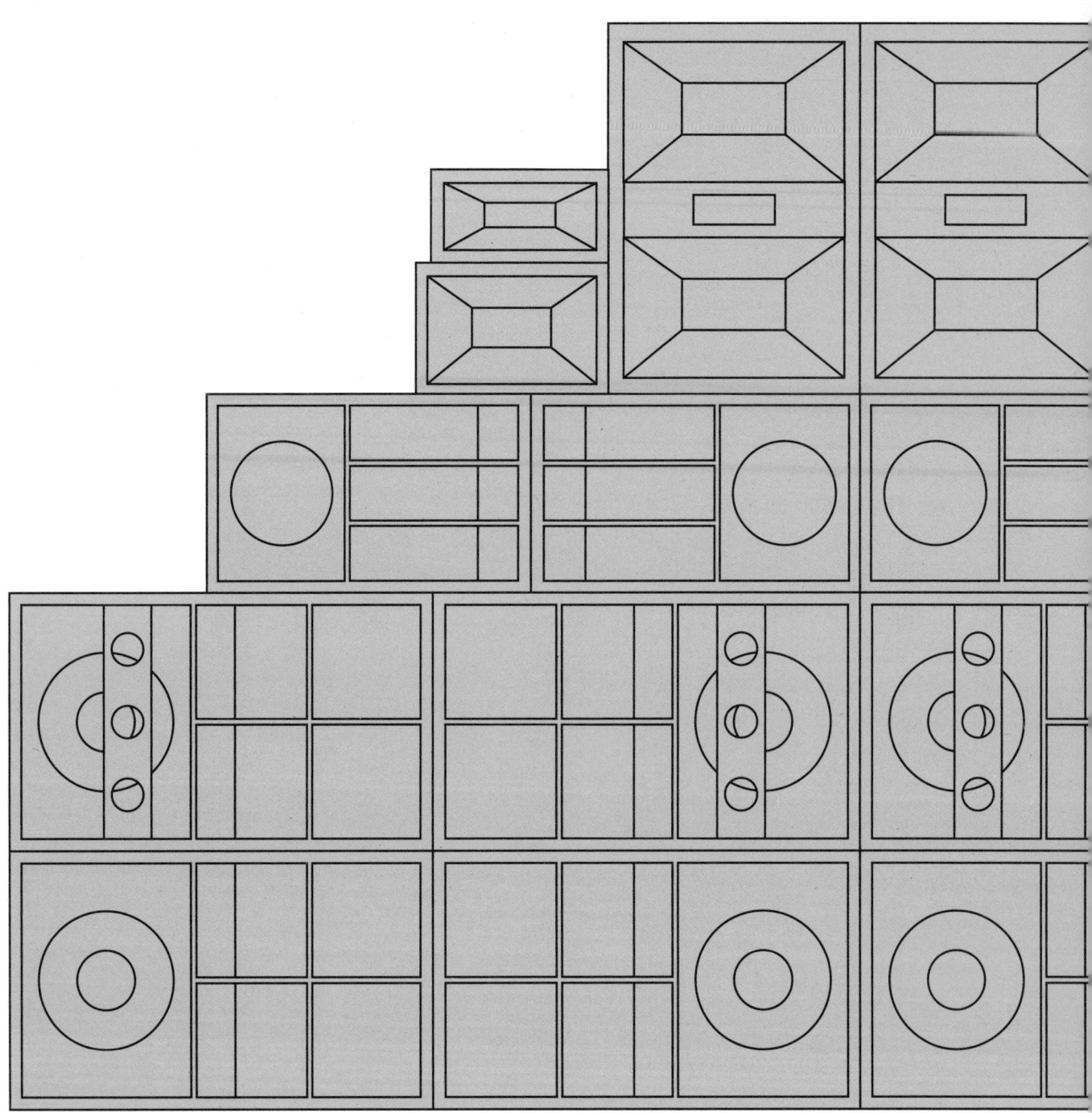

Though barbs had been flung in gossip and song, his teenage music partner Prince Swanny and Rebel were still, in the public eye at least, in the same groove—they appeared on the same playlists, and put differences aside when paths crossed. Thus, early in carnival season 2020, Rebel agreed to be featured on a group recording of Trinibad's key artists, called "Big Badness." It was requested by the island's powers-that-be, to promote peace. With COVID looming and unemployment and fear rising, the murder rates in Port of Spain were raging, and peace was definitely needed. Rebel responded to the call for unity affirmatively.

Carnival offered him a further chance to show support. The government of T&T, like that of all Caribbean islands, has long maintained an official interest in promoting its nation's culture. Trinibad's key figures were put on official alert that they were to disrupt the venerable Soca Monarch Competition, held on February 17, 2020. It was a contest that represented a tradition of friendly creative rivalry quite different from the vicious shoot-outs depicted by today's musicians. T&T's boosters and political establishment were clearly banking on the Trinibad sound to sell internationally, despite or because of its lyrical violence—a changing of the guard

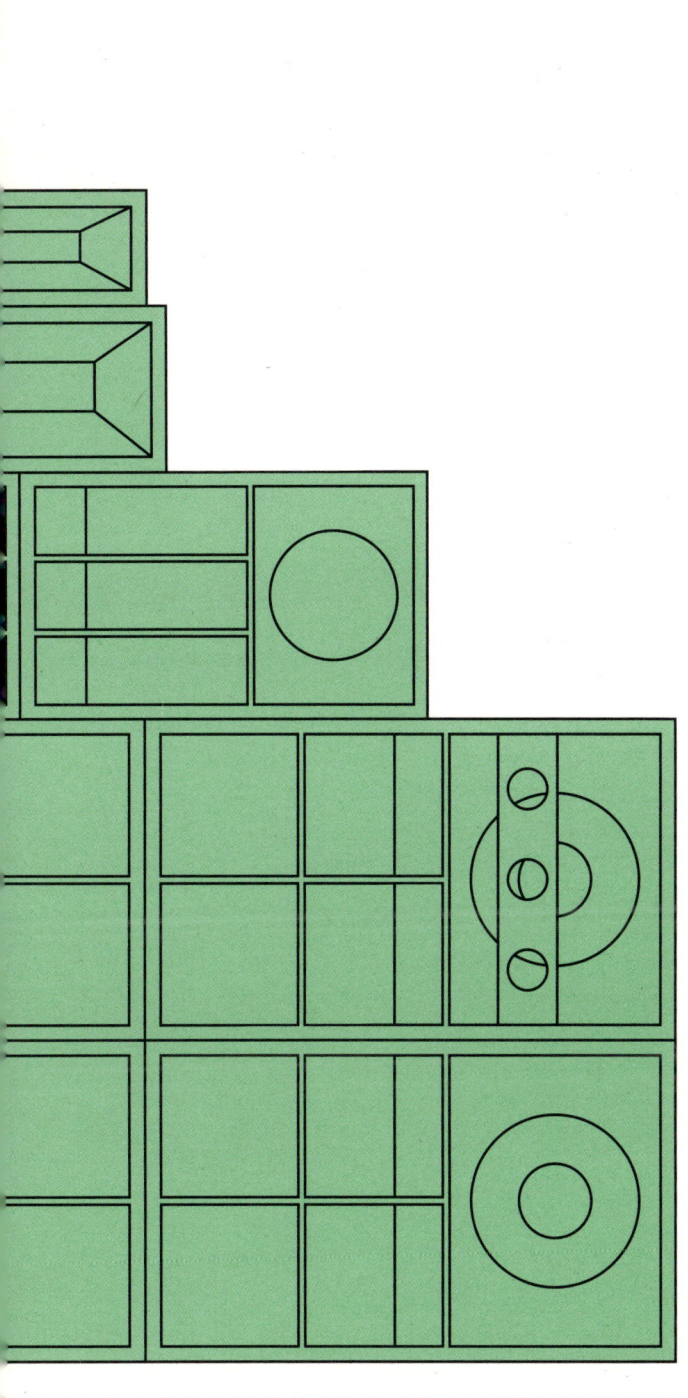

from the feel-good soca days of yore, but still meant, so the invitation said, to promote peace.

Backstage TikTok videos show the bros laughing, drinking, and singing together. But while the talk was of peace, the gang warfare kept going; murder, in many areas, became the norm. To pick up a newspaper in Trinidad was to glimpse images, at least once or twice per week, of a body splayed in the street.

* * *

Right after carnival ended that March, COVID hit the island. Rebel dismissed the "soldiers" patrolling his yard.

It was time to hunker down. But not before Rebel played his first show off the island, in Barbados. He was excited, but still irritated by a problem. When the Trinibad crew played the Soca Monarch Festival in February, they had expected to perform their joint peace tune, "Big Badness." But the release had been delayed. Before leaving for Barbados, Rebel heard that certain artists had never completed their parts and the record had been shelved indefinitely. It was all the more annoying as Rebel had only done it to take one for the establishment team—which was now dissing him.

As Trinidad went into lockdown, Rebel considered staying in Barbados. It was tempting to duck the tensions of home. But he couldn't abandon his folks or newborn son—and then there was the music. The melancholy lure of "No Trust No Love," the first Rebel track to fascinate me, turns out to have been a premeditated strategy on the part of Rebel and Hugh, in tune with the times.

"We in Trinidad had just left a period of peace," Hugh explains. "There was supposed to be no more bad mind. We made 'No Trust No Love' on purpose; our effort to go fully international. I calculated it was a good number one, and so said, so done." On the morning of Wednesday, June 3, after spending all spring locked down at home, Rebel messaged their group chat: "I want to drop 'No Trust' today."

One week after "No Trust No Love" came out, it was top of the T&T Apple Music chart. But then came the blows. On June 10, Rebel and Hugh learned that K-Lion had died in Miami. He was 26 years old and had seemed in perfect health. "It changed Rebel," Hugh said. "They said K-Lion had a heart attack playing football, but it was COVID time and there was no autopsy. It seemed suspicious. He broke down in the car when he heard, and from that day forward, it was a totally different Rebel."

Compounding the loss of K-Lion, just days later Nye—the friendly area leader who was producing a show featuring Rebel—was killed. His fate never made the news. Rebel knew he was trapped. No matter that he was trying, like Marley before him, to stay on good terms with every clique on the island. Whatever move he made would displease someone. By now, the jollity of February's Soca Monarch Competition had evaporated. At a moment when COVID had the island on lockdown, Rebel's success soared and the demands on him mounted. He became mistrustful of the whole music/political/industry complex and threw himself into a frenzy of creativity. Within three months, he laid down 40 new songs—many of which, Hugh recalls, revolved around Bad Mind.

Then came a phone call, from whom Rebel never revealed: an invitation to join a press conference for peace with artists like Swanny, from his old crew, Rasta City, and rival Muslim City crew singers including Toppy Boss. This gathering was one of many responding to what became known as the Morvant Killings—the murder of three men by police in that area of Laventille, on June 27, 2020. Occurring a few weeks after George Floyd's murder in Minneapolis, these killings' impact here was akin to that of Floyd's in the US. As gunfire held citizens hostage nightly, the murder rate spiked alongside mass protests for justice and peace. (In 2022, almost 10 percent of the island's murders occurred in Morvant and Laventille, averaging at least one a week.)

This time, Rebel turned them down. He was tired of playing what he saw as a phony peace game, particularly as he had spent the previous months wasting his time trying to fulfill everyone's expectations. He felt the whole thing was more a cosmetic charade than anything. He had been jerked around about "Big Badness." After the unexpected death of K-Lion and the unacknowledged murder of Nye, he had lost faith in what he now saw as a mere performance of peace. Plus, he was busy. The trajectory of Rebel's career now had a global thrust that looked set to far outstrip Swanny's, or any of his other fellow Trinibad artists.

On July 4, the rival Trinibad crews gathered for peace. Among the artists present were baby-faced Prince Swanny from Rasta City, and "Muslims" like Toppy Boss, whose career was based on singing about the sort of beefs—including one with Swanny, who had one with Rebel—that buzzed about on the many online sources for gossip about Trinibad. Also on hand were Medz Boss and Lawless, Magic and Siah Boss, Bravo and Leo King and Leroy. Bold hopes were stated to journalists, and the mood was loving elation. But where was Rebel Sixx? Everyone knew the missing musician was untouchable when it came to sophistication, subtlety, and depth. In the Trinibad jungle, his was the loudest roar.

The next day, he was assassinated.

* * *

Having been at Bob Marley's side in London and Kingston, in those months and years surrounding the dons' attempt on his life in 1976—and having seen many friends and associates killed after the Peace Concert there two years later—Rebel's murder hit me with a sort of contact PTSD. It seemed, whatever the details, a part of the same sick phenomenon. Bad Mind. A targeting of the most gifted artists that also seemed to share hallmarks—politically motivated gang allegiances—with Bob's narrow escape. I was haunted by Rebel's killing.

I learned that two masked men had broken in at 11:45 PM on Sunday, July 5, 2020, and shot him 20 times at close range at his home in Bon Air Gardens, Arouca, outside Port of Spain; Rebel hadn't heard them enter as he was on his Playstation with headphones on.

No one was ever convicted of the crime, and the police closed the case after a couple of weeks. But it was impossible for the public not to draw its own conclusions. Fans raged that Prince Swanny had been seen on social media happily drinking as if glad of Rebel's demise, but he could have been laughing at fond memories. With virtually everyone in the T&T entertainment industry and associated bodies involved, how could anyone not be implicated?

What failed to feature in Trinidad's articles after Rebel's death was another factor in his complicated, ambitious life. The response to his new song, the 6ix-oriented "868 King," with its steel pans, had been so strong that Rebel was ready to make a change. "Though Rebel would always love Squash, we were trying to leave the 6ix, this stigma and the box that people were already trying to put us in, and become international now," Hugh explains. If there really was love in their bromance, Squash would understand Rebel's need to distance himself professionally—not personally—drop the Sixx, and just call himself Rebel again.

As if to guarantee that some mystery will always envelop Rebel's killing, there is a dispute regarding the date of the peace gathering held at George Street. Sources said it happened on Saturday, July 4, the day before Rebel got shot. Others insist that it took place the previous day, and that the mix-up is a manipulation—misinformation calculated to make people think that Rebel was killed solely because he was not there.

Those around him and Rebel himself were well aware that several factions were after him, hence the planned house move. Like a John le Carré novel, it was a case of spin the bottle to see who—if anyone—succeeded first. Indeed, Rebel had been paranoid, as he understood twisted Bad Mind thinking. He knew that some would blame him for the death of his pal Nye, the music-loving area leader. Equally, Rebel's refusal to co-operate with his successor from a different team might have made a new enemy. Or could a threat come from someone who hated the 6ix, or loved them and heard he might leave? Rebel's reality was fragmenting as if in a hall of distorting mirrors.

Cruelly, the world was just opening up for Rebel. "In the week before he was shot, Rebel got his first international bookings, two in New York and one in the UK," Hugh recalled, as he described making his way to Rebel's house the night he heard of the shooting, only to be told by police the investigation would take its course. "We got a double fee for one of those as a down payment, and after the fact, the promoter said, I don't want it back. Keep that money for the family and funeral."

* * *

The investigation was closed with comparative speed. Maybe we will never know the truth. For Rebel's friends, some connection between that peace no-show and his death does not seem self-evident. Rebel, as his friend and producer El Faltino poetically pointed out, had people gunning for him in high and low places. Instinctively, he turned to the phrase from Ephesians that Bob Marley used, in his song "So Much Things to Say," to describe where wickedness dwells. An optical "peace" mentality is the flip side of that same weighted coin as Bad Mind.

"Every time you see a peace concert, just know that behind it is darkness," opined Hugh. "For somebody to openly propagandize and want to showcase that form of theater, it means they have an agenda."

He recalled other events less grounded in the community than Marley's half a century ago, like the signing of a peace accord between warring gangs in 2005. "In the weeks following, all the guys who signed the Peace dropped, one after the next. When these events do rise authentically, there are bound to be attempts to cut them down from the powers that be. They don't want positive change that doesn't benefit the establishment, as they see it. In no time in history has it been different."

The loss of both K-Lion and his bredren Rebel, so close together, cut the sapling of Trinibad at the roots. Toppy Boss mourned them in a song that was otherwise all bums and guns. But an infinity of mourning can sap energy needed to make change. In the fall of 2023, word emerged of yet another new truce between Rasta City

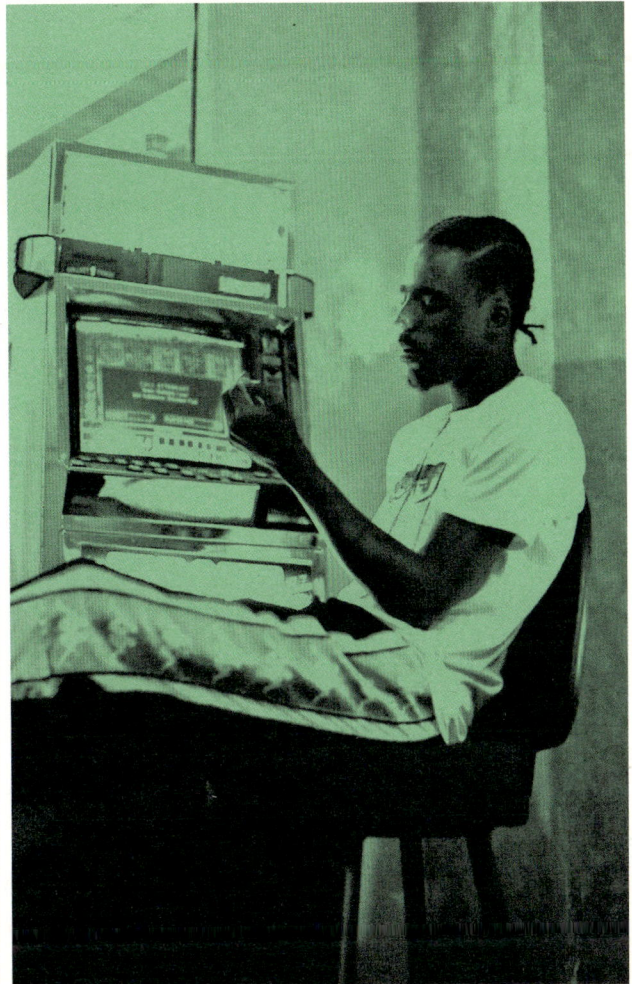

Rebel Sixx. Courtesy of Raya Media

and 6ix. But in a scene so steeped in the nihilism of Bad Mind, is it too late to give *proper* peace a chance?

"Hope is there!" insists Hugh confidently. "Because Trinibad, like rapso, is the new branch of a legacy that comes from steel pan and calypso. They all represent the decision to go against what is popular or deemed acceptable and choose to carve your own voice and your own way. It *is* a decision, and when those guys make that decision, they almost always accept the rest. It is an attempt to try and leave the situation they have grown up in and it is do or die. The irony is, if you do, you might die too, but you might get a better life and be able to get a house for your mum."

No doubt Squash has long since bought a fine residence for his beloved, flamboyant mum, Shelley Anne Millwood. Through his travails, she has become his spokeswoman. As I was driving to the airport to fly home to New York and its own lethal forms of craziness—guns and gangs included—news came on the radio that Millwood had just confirmed Squash's release from those ICE charges in Florida. Interested parties can only await forthcoming music and events from Salt Spring and the 6ix crew, with hope for a sound, soon, that mellows the violence as conscious bass once did. ●

She had a safe word. But she did not want to use it prematurely.

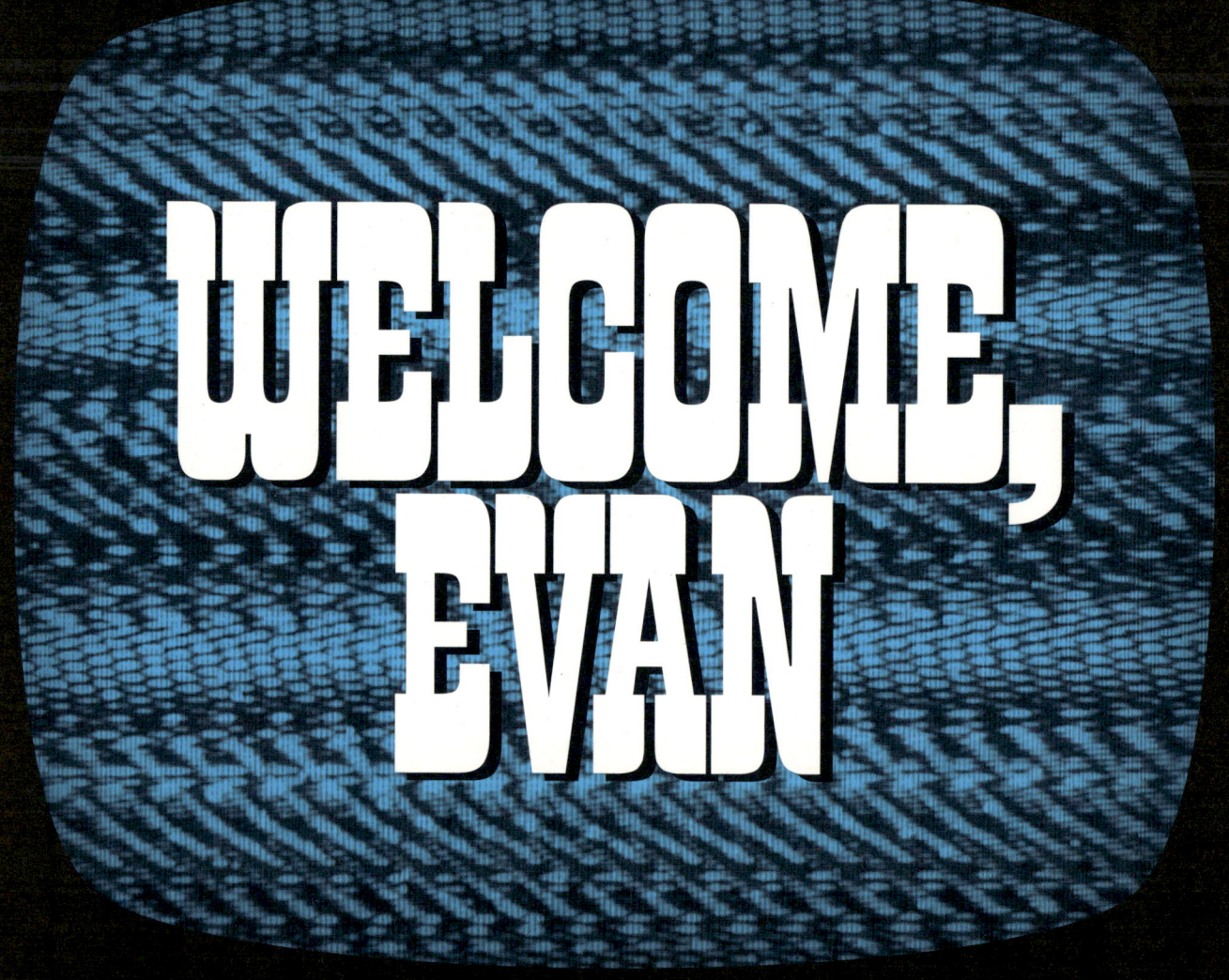

FICTION BY ISLE MCELROY

"Sir, do you need a bag?" the cashier asked as she paid, his words firm like a steel pipe. Edie shook off the cashier. She didn't begrudge the man—how could he know? But she felt a pinprick of shame, for both herself and the man. She didn't want to embarrass anyone; that was often the problem.

Outside, Angela waited in line for a small specialty market that limited entry to two customers at a time. Edie held out the bottles of wine for inspection. Angela was an expert—a sommelier by training, though she didn't like using the term. The people who called themselves sommeliers reminded her of detectives slapping their badges on counters, demanding respect. Edie had been nervous to purchase the wine alone, but Angela insisted she trusted her.

She nodded at the bottles. "Do you know what they're planning to cook?"

Edie didn't.

"These should be fine," she said, and they both broke into laughter. "I really mean it!"

"I know, I know," Edie assured her, and wrapped her loosely in her arms.

A man sprung out the door of the market with his shirt pinched over his nose. He exhaled as if breaking the ocean's surface after a dive. He was thin, unexpectedly bearded under the neck of his T-shirt. Edie hated the man in a way that had become familiar. It was so easy to read malice into forgetfulness—at worst, selfishness, a feeling Edie believed she understood a little too well.

The man looked to Edie for solidarity. "Barely made it out alive," he said.

Edie tightened her mask over her nose.

Rejected, the man moved to a bench. His wife stepped outside carrying a paper sack by its twine handles. She gathered her husband and offered an apologetic nod to everyone waiting.

Edie and Angela bought cheeses and rosemary crackers for the friends they were seeing that evening. Friends. Was that the word for a couple you met on the internet? A couple charging Edie money to see them?

As they walked to the rental car, a white hatchback pulled into the lot, and Edie nearly dropped the wine. Angela asked what was wrong. Edie waited until the driver stepped out of the car before she said, "Nothing." A young woman wearing a baseball cap locked her car with the fob. Her ponytail hung through the back of the cap—Lucy would never have worn her hair in this fashion; the two women looked nothing alike.

"Is that her?" Angela asked.

"I thought it might be," Edie answered. "I'm paranoid for no reason."

"No reason," Angela said. She reminded her to not minimize things.

Edie punched directions into her phone, and over the short drive—not even twenty minutes—night fell like a sheet. The pavement grumbled into dirt until the car was crawling over a dusty, divoted road. "You have arrived," said the phone, as they pulled into an unlit driveway and, as per the email instructions, came to a stop under the car porch.

Edie met Lucy Kay at a writing conference on the Florida coast when she was a man and Lucy was her teacher. She was leading an intensive novel workshop where the students met every morning for one week, three hours a day, to discuss chapters from their unpublished novels. They traded praise and enthusiasm—encouragement was encouraged at the conference.

The first morning, Lucy entered the classroom a few minutes late wearing a baggy, cable knit sweater unfit for the weather and frayed denim shorts that ran a little close to pajamas. "Hello my loves," she said, her tone flirtatious and ethereal. She drank coffee from a small white mug with the name of the conference glossed on the side. Edie had been reading Lucy's work for years and admired her deeply, especially the excerpts from her most recent book that had been appearing online. But in person she seemed wary and needy, like a disgraced pop star unsure how to act when she wasn't on camera.

She was only five years older than Edie but had already published four books to her zero. Until recently, she taught creative writing at Columbia but abruptly quit to "find herself," she told anyone who would ask, though she ignored most follow up questions. Speculations surrounded her departure. She had left *too* quietly, too *quickly*, to have quit on her own terms. Surely a scandal was waiting to surface. Or, perhaps her hasty departure was proof the decision was hers—no hearings or student committees dragging on for months. These kinds of rumors were not new to Lucy. Her first novel teased at her upbringing—her parents had scammed the entire population of a small Colorado town out of their savings—and readers scraped the details of her novels looking for truth underneath, as if they were taking nickels to lottery tickets. She seemed to relish the infamy. Like a porcupine relished its quills.

In class, a landscape description was praised. A mother needed further development. Lucy seemed underinvested and embarrassed to be in the room. It wasn't her first time teaching here, and perhaps she no longer felt the need to prove her value to students. Months later, she would confess to Edie that she wasn't supposed to teach at the conference that year. Another instructor dropped out at the last minute and she was invited as an alternate. "Afterthought" is the word that she used. Once the relationship ended, Edie wondered if *this* was her reason for treating her how she did. Her star had faded and she wanted to matter again, to feel powerful, even if only to one person.

Riley knocked on the driver's window. He was wearing a white disposable mask that resembled the bill of a duck. Edie rolled the window down two inches and accepted the package of rapid tests. "Snap a photo when it's ready," said Riley. He returned to the house to await the results.

"We can always tell them we're positive," Angela said. She mimed drawing a second line on the results box.

"Do I seem that nervous?"

"You don't not seem nervous," she said.

Edie sent Riley a photo of their negative tests. "Come on in!" he texted back.

They lugged their suitcases over the dirt driveway to a pair of sliding glass doors that opened into the kitchen. Riley greeted them with a handshake that clobbered into a hug. He was a few inches taller than Edie and twenty years older, his mop of brown hair smeared with tufts of gray. His beard was a neat layer of puff dyed the color of syrup. His wife, Marguerite, waved from a barstool at the kitchen counter. She had a smooth, plum-like figure; her legs dangled like streamers off the seat of the chair. Her head was buzzed and Edie spotted sun marks darkening through the fuzz of remaining hair. She said, "It's for the wigs," as she stood to greet them.

"What is?" Edie asked.

"The wigs are important," Riley added. "For verisimilitude." He studded every syllable with a little medallion of pride. It was clear he loved presenting the word like a favorite child.

Marguerite beckoned for Angela to follow her, and the two women rolled the suitcases to the guest bedroom. Angela flashed a faux grimace at Edie, as if she were being led to her death.

"Should we—should I pay you?" Edie asked.

"You know this isn't all about money for me," Riley said. "This is my passion. Helping people live their best lives." He opened the freezer and collected a pair of large round ice cubes and dropped each one into its own squat glass. "Gin and tonic okay?"

"We brought wine," Edie said, and bent down to retrieve it.

"We'll do that with dinner."

"Gin and tonic sounds great."

Riley nodded at a thin stack of papers on the counter. "Flip through that and see if it's in order. We're planning to start with the beach in the first trailer, couch in the second, then the bed, and end with the walk to the bar. Four total. Does that sound right?"

Dread thickened to the edges of Edie's body. She glanced at her fingers—a grounding technique taught to her by a former therapist—and focused on her nails, painted road sign orange that morning, to match the desert sun. "I think so."

"You think so or you know? This isn't horseshoes. We've gotta be one hundred percent."

"I'm sure," Edie said. "Sorry. I'm nervous."

"Nervous for what? You're about to get better." Riley winked and passed her a glass. He tipped his forward to clink. Edie held in a cough tasting the heft of liquor that Riley had given her; she was relieved to blunt her nerves. "You sign on the last page. And put someone other than your wife for the emergency contact—not much use if she's here."

"She's not my wife," Edie said.

"You know what I mean."

"It's been about a year," she said. "She's been great about everything. Compared to—"

"Compare and despair my friend."

"Thirty-six hundred?" she asked.

"Four thousand. With the room."

"Of course—you said that." Edie didn't remember him saying that, but there was no point arguing after coming all this way, after having already obtained so much money in cash. She squatted to retrieve the envelope out of her backpack pocket. Inside was only thirty-six hundred. "You don't have an A-T—god of course not."

"There's one in town. You can pick up the rest after your hike."

Edie thanked him, then passed the envelope over.

"Normally I'd call this whole thing off, you know. It's not worth it for me if I can't feel like I trust the person, and money—I hate to say it—is the fastest way to build trust in this world. Don't think I like it. I won't say that. But sometimes it is how it is."

"Should we find another place to stay?"

"What I'm saying is normally I need money to trust a person. But you're different. There's something about you."

"People don't normally take me as trustworthy," Edie said.

"I'm not saying you are," Riley said. "But you're scared. You're scared of me. Which is weird because I'm not a scary person. I'm very nice. Ask Em. Em, am I a scary man?"

"The scariest," Marguerite said, as she and Angela returned to the kitchen.

"Oh she's biased," he said.

Angela came up behind Edie and wrapped her arm around her waist. She sniffed at the gin and tonic, took the glass from her hand. "I'm not scared of you," Edie said.

"This is a site of honesty, Edie."

"Maybe a little," she said.

"There we go," Riley said with a laugh. "Much better."

A few weeks before they broke up, Edie and Lucy flew to Joshua Tree to check on a house Lucy wanted to buy. It was November, the desert cool and polite; daylight was pinched tightly between sunup and down. Lucy had wanted a house in the desert for years, and after the sale of her latest novel—her fourth—she had the money to buy one outright. It was a squat stucco ranch without running water; in the photos online, dirt showed through cracks in the floor. It would take months of remodeling

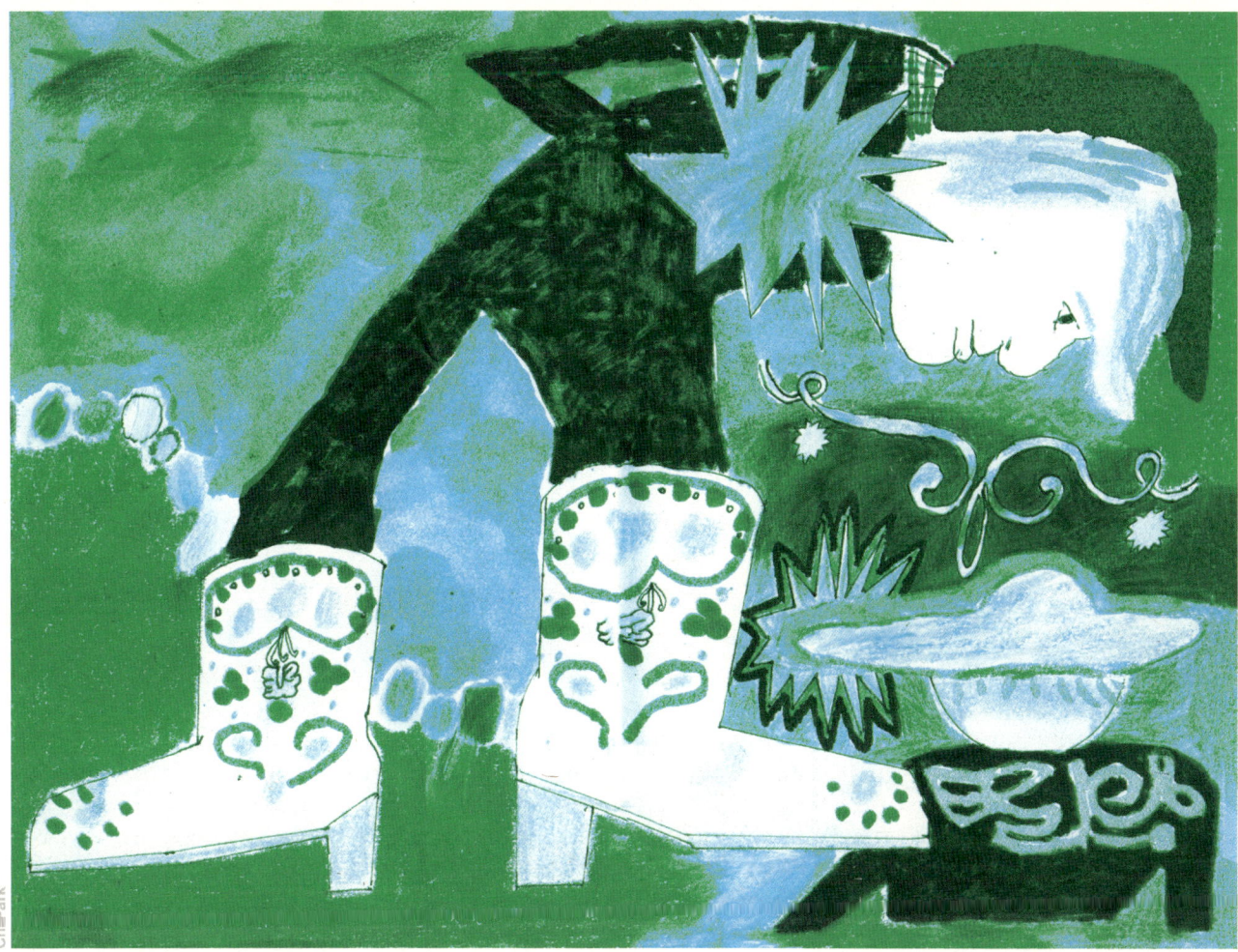

Chris Park

before it felt like a home. It was not the most elaborate or stylish house, and it wasn't supposed to be. It was a house where writers could live, cheaply, writing their books.

"What do you think?" Lucy asked when they parked at the end of the driveway. They both got out of the car and leaned against the passenger side.

The current residents weren't home to let them tour the inside. Edie squinted to get a look at the house from the road. "It looks like the pictures," she said.

"See that covered porch in the back? You can put a table out there and spend your mornings writing outside in the shade. We'll hike in the afternoons. You can do whatever you want. There's not much water—we won't be showering much. The sex will be sticky and gross." She laced her fingers through Edie's. "No phones. No Twitter. No stupid Internet fights and dumb shows on

TV. Just the landscape and us and our books. Everything you wanted."

Lucy liked to remind Edie what she wanted. What she wanted, Lucy insisted, was a long career writing books, and, living with Lucy, she could write as much as she wished. It was the perfect opportunity for her to build a career. Edie wouldn't have to worry about money. Lucy would pay for everything. She would have to—Edie was down to a few hundred dollars after failing to find work in Denver, where she had moved to live with Lucy.

Lucy already paid for meals and groceries and trips to the movies and had even bought Edie a new bike after hers was stolen. Edie had a bad habit of allowing arrangements like this. Whenever a friend offered to buy a round, she accepted; she never argued when told not to worry about paying someone back. This was how Edie had always existed, beneath the

circling palm of others' beneficence. It seemed foolish to refuse generosity. Now, though, she worried what might happen if Lucy changed her mind. Would she really let her live there forever? Did she even want to?

"How does that sound?" Lucy asked.

"That sounds wonderful," Edie said, but she angled away from Lucy. After only a few months in Denver, she felt tethered and possessed, even as Lucy encouraged her to branch out into the world and discover who she wanted to be. She was no longer confused about who she wanted to be—at least, not as confused as she'd been after coming out—but she was scared to leave the safety of knowing for the reality of experience.

She wasn't an idiot. She knew the risks of pursuing the life she desired, and had already lost so much. People in her life loved reminding her of the things she would lose, as if it never occurred to her. She kept a running tally of what she would lose: stability, finances, pickup basketball, how handsome her jawline looked beneath a fine mist of stubble, safety.

Edie moved across the country to live with Lucy because she longed to live authentically. In Denver, in the home Lucy was renting, Lucy had given Edie makeup and old dresses and nail polish and encouraged her to dress as herself in her home, but that freedom had begun to brush against obligation. Lucy presented herself as a caring and safe person. However, standing beside her surveying the ranch house in the desert, the house they could very well share for the rest of their lives, it occurred to Edie that she hadn't felt safe since Lucy entered her life.

That evening, at their cabin, Lucy was drunk. Edie lounged on the couch wearing a black cotton dress and sheer tights. She had shaved her legs that evening—partly because she wanted time away from Lucy and could get it by taking a bath—and kept rubbing her knees together, pleased by the sensation. Lucy squatted in front of the fireplace drinking a cooled can of beer. It was her fourth, so Edie had switched to water. The fourth drink always fractured something in Lucy, and her anger emerged more easily, like cold air through a broken window. Edie had learned to drift back toward sobriety in these situations, should they stumble into a fight.

"Maybe I'm just a way station for you," said Lucy.

"You can't really mean that."

"I'm a stop on your journey. We can't be everything for everyone—I'm not deluded."

"That doesn't make me feel better," Edie said.

"I'm not trying to make you feel better," she said. "I'm trying to make *me* feel better. I brought you all the way out here, I'm offering you whatever you want, and you can't even say you'll come with me. What more do you need?"

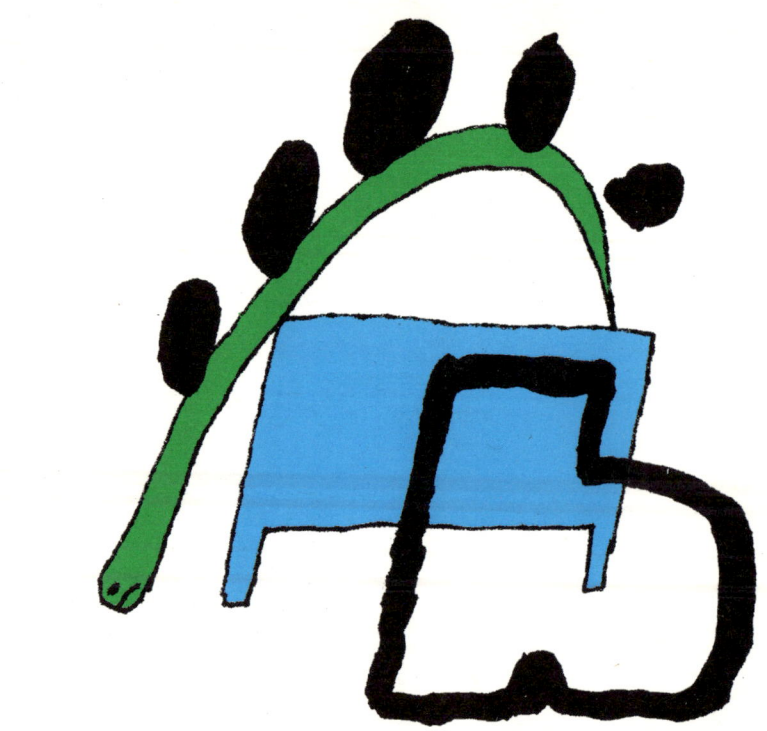

Eric Reh

Edie apologized. Lucy was right—she had given her everything she had wanted and had asked for so little in return. The dress she was wearing belonged to Lucy, so did the eyeliner—Lucy even applied it—and she ought to be grateful. Wasn't this the life she wanted? A world where she could write unimpeded? Where she could be her true self?

Lucy stepped to the couch and stood over Edie. "I just want you to appreciate all that I'm doing for you," she said.

"I do," Edie assured her.

"Then you need to show me," she said.

Edie unbuttoned her jeans.

Angela and Edie left for the hike shortly after sunrise. Angela planned the excursion using a dusty guidebook she found in the bedside table in their guest bedroom. The grayscale maps inside appeared to have been drawn by hand and photocopied haphazardly, pitched onto the pages at troubling angles. At the trailhead, Edie checked the publication date. The book was more than thirty years old.

"Thirty years is nothing in geological time," Angela assured her. "The boulders haven't moved. The dirt didn't blow to the other side of the desert."

"Okay, okay," Edie said, but it was too late to stop her.

"The cactuses are still in the same positions." She stretched her arms to the side, one bent up at the elbow, the other bent down, her face frozen in a cactus' gaze.

Edie kissed her on the cheek, a white flag of a kiss, and Angela laughed victoriously.

Last night, they'd finished the wine and nearly all of the gin, and now their legs were heavy, faces puffed, their words emerging without precision. Edie didn't own the appropriate clothes—she had never owned hiking clothes—and the closest approximation she found in her closet, before leaving New York, was a pair of black running pants and a blue sweat-wicking shirt, both of which had been gifted to her by Lucy. So much of the present seemed to belong to Lucy, as if every part of Edie's life had been rented from her and could be repossessed when she least expected. Edie was here, though, in the desert shelling out four thousand dollars, plus the cost of plane tickets–hers and Angela's–and a rental car, to gain some control over the present. She wanted to be in the desert with Angela, her partner of nearly a year, and not trapped in a grain silo of memories, sinking ever deeper the more she tried to climb out. But she feared it might be impossible to ever get out, that the best she would ever accomplish was not escape but dragging another person into the silo beside her.

The terrain was flat, the path squeezed tightly between sage brush and the extra-terrestrial limbs of Joshua trees. Even at eight in the morning, the air was unforgiving and dry. Edie worried they were drinking too much water, too early, but Angela insisted it was best to drink when you were thirsty, without worrying how much they had left—they had more than enough.

In her early 20s, after college, Angela had worked as a trail guide in Sedona. She took the job on a whim because she wanted to get as far away as possible from Cambridge, where she had sat in a series of smaller and smaller rooms over the course of four years in order to obtain an economics degree she found arbitrary and suffocating. She wanted to be irresponsible, and moving to the southwest, where she knew no one, to pursue a career for which she had little experience, was the least responsible thing she could do after college.

She lived in the desert for two years, the only Asian trail guide in Sedona, she joked to her friends when they asked how she liked it, because she wasn't sure whether she liked it. Liking it seemed beside the point. She needed a break from the path she had put herself on, and there was something fitting about finding herself on a series of literal paths, from sunup to down, pointing out landmarks and rattlesnake holes to the kind of people she had nearly become.

The sky was a wide, breathless blue, and by 10 Edie was soaked in hangover stink.

"It's not much farther," Angela said.

"What is?" Edie asked.

"The Wonderland of Rocks," Angela said. "I told you a thousand times." She didn't get angry often, so it was easy for Edie to recognize when she was.

"My mind is elsewhere," she said.

"We should eat something," Angela said. They rested on the smoothed foot of a boulder, sliding down as they shared handfuls of nuts. She said, "I can't keep waiting for you."

"Waiting for me to what?"

"You know."

"To move on from being assaulted?"

"It's unfair when you say it like that."

Edie knew she was right.

"We're in a relationship," Angela said. "Or we should be, but I never feel like you're with me. You're always in some other conversation or some other moment, talking to memories, ghosts, whatever you want to call them, while I'm right here, trying to have a life with you, and you're not anywhere in it."

"I'm in it with you right now."

"Where are we going?" she asked.

"We're on a hike."

"Tell me the name of the place where we're going. The place I keep bringing up."

"I'm bad with names," Edie said.

"No you're not," Angela said.

"The Rock House."

Angela took a long drink of water. "I am a very patient person. If I weren't, I would have ended this months ago."

"I guess I should be more grateful," Edie snapped.

"But I love you. That's why I'm patient. I'm not trying to threaten you or make you feel bad, but you keep tossing me aside for people who haven't been in your life for three years."

"Why come out here with me then if that's how you feel?"

"Because I want to support you. I want this to work—whatever *this* is, whatever you're paying that kook to perform. I don't want to start over with someone else. But if I have to—I will. I want you to know that."

Edie was ashamed for not seeing this sooner. Would she have put up with similar treatment? That wasn't the point. The point was that she'd never know—she was always the one to have drama. She was a heavy person, emotionally, and she regretted making Angela carry the weight she had placed on her. "How much longer to the Wonderland?" Edie asked.

"Bad with names, huh?" Angela laughed.

Edie kissed her chastely on the mouth, their lips too dry for anything more. They packed their bags and continued.

"Where'd you get all this sand?" Edie asked.

"That's a Riley problem," said Riley. "You need to focus on Edie problems."

Her problems included getting comfortable in the trailer, where Riley had piled a few hundred pounds of play sand to create the impression of a beach. Marguerite was sitting beside her wearing a green one piece like the one Lucy had worn the day after the conference, when she brought her to her friend's beach house. The color was off—Lucy's suit was a richer, shinier green, and there had been frills over the waist—but it was close enough to matter, and Edie was impressed by the care they had taken in trying to get it right. The wig Marguerite wore was uncannily accurate, the exact length of Lucy's rib-length hair, its tangled, yellow-cake blond.

As Riley set up the projector, Edie took a stab at small talk, but Marguerite didn't respond. She inserted a small black earpiece into her left ear so that Riley could feed her her lines. Angela watched from outside the trailer with her arms crossed.

Riley flicked on the projector and a blue blanket of light coated the wall. He connected his phone, counted to three, and the blue transformed into rippling waves spotted with children and parents flouncing around in the water.

"I'll leave you two to it," Riley said.

Angela rushed in to give Edie a hug. "Love you," she said.

Riley slammed the door behind him, and Edie and Marguerite were alone.

"It's so beautiful here," Marguerite said. She angled a little closer on the towel they were sharing. The scene had begun. "Do you want another one?"

"Sure," Edie said. She crossed her arms over her bare chest, still so flat and dotted with the few stray hairs she hadn't caught shaving that morning. She planned to start taking hormones when she returned to New York, after completing the exercise. Starting estrogen before her trip to the desert, she feared, might hinder her chances of obtaining closure. Riley had said nothing to make her believe this, nor had any of her friends back in New York, but she maintained an unwavering faith in this notion, out of fear and intuition alone, as if she were living 4,000 years in the past and terrified of enraging the sun.

Edie's body was slender and firm as a diving board, and that afternoon she wore the same swimming trunks she had worn to the beach and no shirt. I'm topless, she thought, which she hadn't thought the day Lucy brought her to the beach, so she tried to put that thought—that feeling—out of her mind, in service of the exercise.

Marguerite opened a beer and passed it to Edie, in such a way that their fingers grazed during the exchange. She took a sip and set the can to her left. Marguerite tilted her knee so it rested against Edie's. "We didn't touch," she whispered. Marguerite removed her knee without speaking.

"It's getting pretty late," Edie said.

"We're nearly out of beer," Marguerite said. She opened a square mint tin and pulled out a joint. "Do you want to split this?"

"How many have you had?" Edie asked.

"This is our third," she replied.

"Will you be able to drive?" she asked. "Back to my hotel?" Edie only agreed to join Lucy at the beach on the condition she take her back to her hotel. Though *agreed* is too strong of a word for what happened. The day after the conference, following a late night of flirting, Lucy texted Edie to ask if she wanted to go to the water. There was a riverfront downtown, home to a stretch of restaurants and shops, and Edie assumed that is what Lucy meant by "the water." In the car, Lucy passed the exits for downtown, and when Edie asked where they were going, she said, "To the beach. I'm housesitting my friend's place on the water. It's stunning out there." She promised her she would return her to her hotel that evening, and Edie accepted this arrangement because she liked Lucy, and she liked that Lucy liked her, and, she would later have trouble admitting, she wanted things from Lucy, things like professional advice and book edits and praise for the novel she was currently writing. Lucy had agreed to put in a good word with her editor. Edie feared Lucy might revoke this promise should she make a fuss.

"I was thinking about your hotel," Marguerite said.

"It's an hour drive," Edie said.

"I really need one more joint," Marguerite said. "And I'm not sure I'll be able to drive all that way if we smoke it." She was reading from a script—the very same script that Edie had provided to Marguerite—but Edie was rattled to hear these words, words she'd replayed in her head in the same particular voice for three years, from a stranger.

"But I need to get back tonight."

"Do you really?"

"I'd like to, yes."

"So you'd *like* to, but you don't need to."

"I paid a lot of money for that room."

"Have you ever heard of sunk cost?"

At this point, three years ago, Edie had deflated. She'd agreed to stay on the condition that she sleep in the guest bedroom—only after Lucy sweetened the deal by offering to buy her dinner, saving her another thirty dollars she would have wasted ordering room service.

"I'm not interested in what you want from me," Edie told Marguerite.

"And what do you think I want from you?" she asked.

Edie was aroused, and covered herself with her left hand.

Marguerite noticed. "Are you sure you don't want anything from me?"

There was a time, after Edie ended things with Lucy, after she moved back home to live with her parents, when she would spend entire nights, midnight until six in the morning, pacing around her childhood home replaying this scene on the beach. As she paced, she repeated, "You promised to take me to my hotel. Take me to my hotel." She attacked herself for not saying something so simple and true to Lucy when she had the chance. How much better her life would have been, she believed, if only she'd stood up to Lucy; she was ashamed of relenting, because it seemed to imply that she wanted this, that Lucy had seen something inside of her she failed—or refused—to admit to herself.

"Okay," Edie said.

Riley pried open the door.

"That's it?" she asked.

"This isn't easy," he said.

"You were so good," she said to Marguerite.

"That lady's a true professional," Riley said.

"You're doing amazing," said Angela, back at the entrance.

Riley heaved the door closed.

The second time, Edie told Marguerite what it was she wasn't interested in doing with her. "I don't want to hook up with you," she'd said, her voice as earnest and small as a suc-culent, and Marguerite had replied with laughing indignity.

"That's what you think of me? I'm your teacher. That's wildly inappro-priate of you to even suggest it. My god. This is my livelihood. This is my career. Do you really think I'm the type of person to trade sex for pro-fessional favors? One of those bald-ing tenure-track creeps preying on freshmen. How old are you? Thir-ty? You're not a child, okay, and if I wanted to have sex with you—I can't even believe you would accuse me of that—I would ask you, like an adult, because I'm an adult and we're both adults. I wouldn't play whatever ma-nipulative trick you're insisting I'm playing on you."

Edie apologized. They started the scene over again. The third time, she gave up as soon as Marguerite handed over the beer. The fourth time, she relented when asked to split a joint.

"Have you ever heard of sunk cost?" Marguerite asked for the fifth time.

"I have," Edie said. Her exhaus-tion had mutated into anger.

"So you know it'll be easier if you stay. We can hang out more, have dinner together—on me. How does that sound?"

"It sounds like you're a predator," Edie said.

"I'm sorry?"

"A predator."

"You do know what I've been through, right? What I've done for women—for victims."

"I saw the stories," Edie said.

"I bought a billboard—out of my own pocket. *Stop Rape*, right outside the Stanford campus, a week after that predator—that *real* predator—got off with a slap on the wrist." Mar-guerite's performance—the sly mix of disbelief and aggression—no lon-ger impressed Edie; instead, it dis-turbed her how easily she fell into this persona. "And who are you to accuse me of that? Predation. You're just some student—I brought you

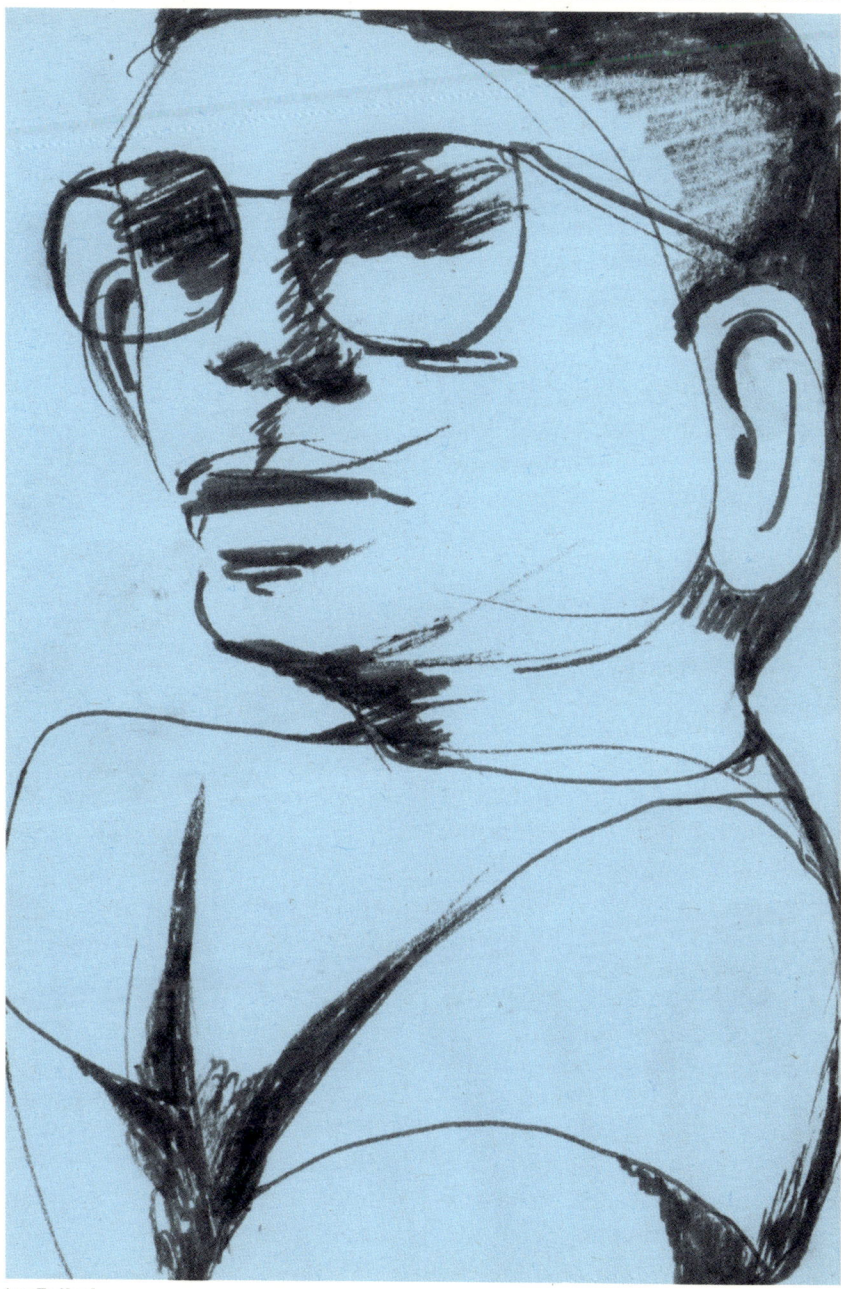
Luc Tedford

here because I feel sorry for you, because you have talent, but you—like so many 'writers' your age are so obsessed with bringing down the people who paved the way for you to have a career. It's sad."

"You're sad," Edie said. "You're pitiful. You're an accomplished writ-er. You've published four books. You've won awards. You taught at Columbia."

"Everyone teaches at Columbia."

"And you're so insecure you hit on your students, the people you prom-ise to help, because they're the only people who don't yet see you for who

you are. It's sick. It's humiliating. I don't want any part of it."

For three years, Edie imagined bursting into tears had she said this to Lucy, the very speech she had practically liquefied through countless iterations, but when Ri-ley pulled open the door, clapping for Edie, she felt a bursting, unsat-isfied rage she had never known in her life. She wanted to keep scream-ing at Marguerite, to sand her down with misplaced resentment, and, per-haps sensing this, Angela came to her and stepped in front of her rage like a mother rushing between a car and her child. "I'm so proud of you," she said. "Bring that to the next one."

In the next trailer, which had been fashioned to look like Lucy's living room inside the home she owned in Denver, Marguerite straddled Edie on the couch. They were both in their underwear, as stipulated in the contract, and when Marguerite lifted her left hand to Edie's throat, tightening her thumb and middle finger around the base of her neck, Edie slapped her hand away. "What are you doing?" she asked.

"I think it's hot. You don't think it's hot?"

"No, I don't think it's hot when you choke me without asking permission."

"Can I choke you?" she asked, in a sheepish, ironic voice.

"You can't," Edie said.

That was it. Riley entered the trailer. "Record time," he said. "I don't think I've ever seen someone get through it so quickly. You can feel how it's working, right?"

Edie nodded. She knew he was praising himself, not her, and that endorsing his praise of himself was built into the cost.

In the third trailer, Edie and Marguerite lay together in a king-sized bed, the edges of which pressed against the walls. "Goodnight," Edie said, and pretended to turn off a lamp. Marguerite shut off her non-existent lamp and scooted closer to her under the covers. She wrapped her top arm around Edie's chest and lowered her hand to her waist while kissing the back of her neck.

"I'm too tired," Edie said. Marguerite slid her thumb under the waistband of her boxers—careful to not touch her where they had agreed to avoid but was close enough to make Edie aroused and frightened. "I'm tired," she repeated. But Marguerite continued, guiding her fingers delicately along the inner edge of her thigh. Edie wanted her then, and was repulsed by wanting her, this imitation of Lucy doing the very things she despised Lucy for doing. Edie

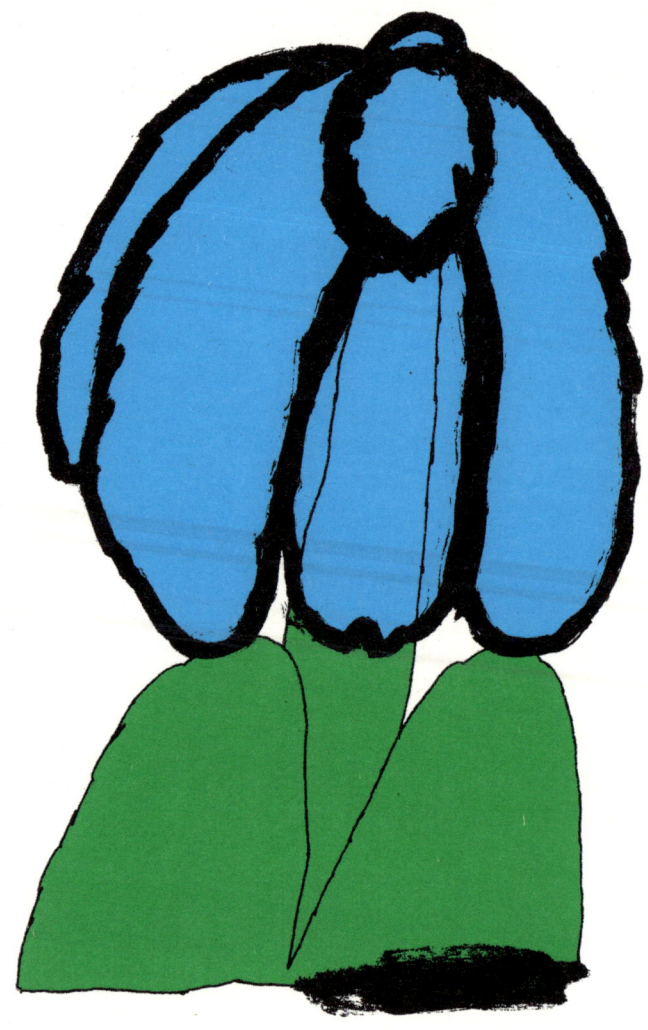

Eric Reh

wrenched Marguerite's hand out of her boxers. "What don't you understand about I'm too tired?"

"You're tired?" Marguerite said. "You should've told me."

"You heard me."

"You're always so quiet, you're practically whispering. You need to tell me when you don't want to."

"I told you, and I'm telling you now."

"Are you sure?"

"I'm sure," Edie said. "I'm very sure that I don't want to be raped."

"Please," she said. "Don't just throw around words."

"Go sleep on the couch," Edie said. It was more than she'd planned on saying.

"Excuse me."

"I won't sleep in the same bed as you."

"You know this is my house, right?"

"If I go downstairs I'm calling the cops."

"So you're blackmailing me?"

"I'm protecting myself," Edie said, taking more pleasure in saying this than she expected. "You have ten seconds."

"You're a psycho," Marguerite said.

"Nine seconds. Eight. Seven."

Riley slid open the door. "Goddamn! You two are like ..." He made a chef's kiss gesture. "Impeccable chemistry."

In the fourth trailer, Marguerite and Edie paced from end to end, simulating a walk longer than the space allowed. The walls had been painted to resemble the restaurants and stores that Edie and Lucy had passed the first night she went out dressed as a woman.

Edie still owned the dress she had worn that evening—a slim, knee-length dress the color of black cherries—and was wearing it now, in the trailer. Lucy let her keep it after she left.

Marguerite asked Edie how she felt to be in the world this way, and she told her she was excited but scared. They continued speaking without saying much, passing time until Marguerite said, "You know, I was talking to my friends last weekend, everyone who came over, talking sex stuff, and they were getting so bent out of shape about consent, and I told them I don't even believe in verbal consent. I think it's all so over-determined. I said I do stuff to you all the time without asking, and afterward we decide whether it was okay. I think that's how it should be. But they acted like I was some kind of freak."

"Is that what you want? To do whatever you want to me?" Edie asked.

"Of course not," Marguerite said. There was a tremor in her voice. She pulled her hair back, nervous, and knocked her earpiece loose to the floor.

"Without asking?" Edie said.

"What's the phrase—better to ask for forgiveness than permission."

"I think it's the other way around."

"There are things that I want," Marguerite said. "There are things that everyone wants."

Edie wanted this conversation to end. And she could end it—Riley had made that very clear to her, during orientation. She had a safe word. But she did not want to use it prematurely, not when she was already doing so well and was so close to the end.

"I want you when I want you," Marguerite said.

"That's just not possible."

"You get me when you want me."

"What are you saying?"

Marguerite crouched to retrieve the earpiece but accidentally kicked it into the corner.

"You assaulted me, Lucy," Edie said. "Multiple times."

"So now you're using the A-word? Is that the story you're telling yourself?"

Edie spied the earpiece and wondered whether Marguerite was wearing a second one. She no longer sounded like Lucy—or not entirely like Lucy—but a blend of the two women. Edie couldn't tell whether she was imitating Lucy or speaking as herself.

"Do you know the types of things that I've been subjected to? What happened to my mother? To my grandmother? To every single woman in my life? And you're acting crazy because I accidentally grazed your neck during sex."

"It wasn't an accident."

"You could kill me, you know."

"What does that mean?"

"If you wanted to."

"Marguerite?"

"It would be so easy for you. You're so much stronger than I am."

"Please, Marguerite."

"It's what men do when they're angry. They kill their girlfriends. They *choke* them. The women never choke back. I dare you—tell me one story of a woman who chokes a man back. So don't pretend like you're scared of me when I'm the one who should be scared of you."

Edie didn't realize she'd pushed Marguerite until the woman was falling. She lunged to grab her, as if she might yank her back into a stand, but by the time her hand was lowered Marguerite was already on the ground, in a fetal position, protecting herself.

"You back the fuck up," said Riley. Edie hadn't heard him open the door. He crouched over Marguerite and removed her wig, inspecting her head for abrasions. "You're not bleeding," he told her. She landed on her backside and braced the fall with her forearm; as he examined her, she turned her wrist in circles to lessen the pain.

Edie let out waves of apologies, not one of which was acknowledged.

"You should already be gone," Riley said. "And if you even look at me like you want a refund I'll sue you to the end of the earth."

At the nearest hotel, Angela requested a room with two double beds.

"Should we push them together?" Edie joked, as they set down their stuff.

"I'm gonna take a shower." Angela disappeared into the bathroom.

As the tub faucet ran, Edie listened for the sound of the showerhead screaming. What she heard was water splashing in water. Angela was taking a bath. The faucet stopped running. Water sloshed as Angela slipped into the tub.

Edie flicked on the TV and mindlessly climbed the channels, passing home remodeling shows and trash pickers scraping through repossessed barns and chefs competing for prizes to pay off medical debt. She clicked until the channels reset to the hotel's directory page. Gentle piano music played over a slideshow of local attractions. At the bottom of the screen, in large bold script, were the words *Welcome, Evan*. Edie shut off the TV.

Angela unplugged the drain, earlier than expected, but remained in the bathroom. Edie tapped her knuckles against the door and entered before Angela answered. She lay naked in the empty tub. Her hands were over her face, knees bent at a soft angle. "Are you okay?" Edie asked. Angela inched to the far wall to make room. Edie eased herself down beside Angela, still wearing the loose T-shirt and mesh shorts she put on to sleep. She kept one leg spidered over the rim of the tub, foot flat on the tile. Eventually, she fit her arm under Angela's waist, and Angela draped herself over Edie's stomach, head resting on her neck. They remained this way, until the discomfort became too much to bear. ●

Picture This: The Periodic Table

PHILIP BALL

On the visual ordering of all the elements known to humankind—and how we might order them differently.

It looms over every chemistry classroom and lecture theatre, two towers bookending serried ranks of compartments rather like the British Houses of Parliament—and with at least as much authority. The Periodic Table is chemistry's icon, a codification of the entire chemical universe expressing the relationships between the elements from which all ordinary matter is constituted. There are just 92 or so of these natural elements (occasional oddities like technetium, element 43, barely exist in nature because they are so unstable), but scientists have now appended to the roster a gaggle of extra ones, from neptunium (element 93) to oganesson (118), that are wrought artificially in nuclear reactions, the most massive of them living for just an instant before decaying.

I must once have known the Periodic Table by heart. It's too long ago for me to be sure, but I have to surmise as much because chemistry students at Oxford University weren't given the table for their exams. In a gesture of characteristically perverse exceptionalism, we were expected to memorize it. But don't ask me now where to locate rhenium or iridium: all those obscure transition metals in the long central block of the table are a blur. If, however, you do know

where an element goes—if you can assign it to the right row and column of the table—you can deduce a lot about it. You can figure out how the electrons in its atoms are arranged into shells, and make good guesses about the types of compounds it forms, its melting and boiling points and its propensity to react with other elements. In my finals, I wrote an entire essay about niobium—niobium!—on that basis. Goodness knows what it said.

The Periodic Table was conceived as a scheme for bringing order to the elements. When there were deemed to be only four of these—the earth, air, fire, and water of the Greek philosopher Empedocles (it was just one of the elemental systems proposed in ancient times, but enjoyed the weighty advocacy of Plato and Aristotle)—things seemed simple enough. But during the Renaissance, natural philosophers were increasingly forced to accept that the metals then known—copper, iron, lead, tin, mercury, silver, and gold—were not as interconvertible as the alchemists believed, but seemed to have an elemental primacy about them, too. More and more of these became recognized—zinc, bismuth, cobalt, and others—along with other new elements such as sulfur, phosphorus,

carbon, and, in the late eighteenth century, gaseous elements like nitrogen, hydrogen, and oxygen. When the French chemist Antoine Lavoisier (who named those latter two) drew up a list of known elements for his seminal textbook *Traité élémentaire de chemie* in 1789, he counted 33—including light and heat, which he called caloric.

The list didn't seem to be arbitrary though. In the early nineteenth century, several scientists noted that some elements seemed to come in families, resembling one another in the kinds of reactions they engaged in and the compounds they formed. Some claimed to see triads: the halogens chlorine, bromine, and iodine, for example, or the reactive metals sodium, potassium (both discovered by English chemist Humphry Davy in 1807), and lithium (identified in 1817). Was there a hidden pattern to the elements?

The Russian chemist Dmitri Mendeleev, working at Saint Petersburg University, is usually credited with discovering that pattern. A Siberian by birth, with Rasputin-like dishevelled hair and an irascible manner, he published his first Periodic Table in 1869. It is "periodic" because, if you list the elements in order of their mass, certain chemical properties seem to recur periodically along the list. The table is produced by folding that linear list so that elements with shared properties sit in vertical columns (although Mendeleev's first table had them instead in rows, effectively turning today's table on its side).

Mendeleev's insight wasn't unique; by then the existence of a periodic structure that organized the known elements was clear to others too. In particular, the German chemist Julius Lothar Meyer drew up a table almost identical to Mendeleev's in 1868, but he didn't get it published until later—and so missed out on the accolades, to his immense chagrin. Mendeleev, however, had the foresight to see that his table only worked

A variation of the standard periodic table, as of 1975, by James Franklin Hyde.

96

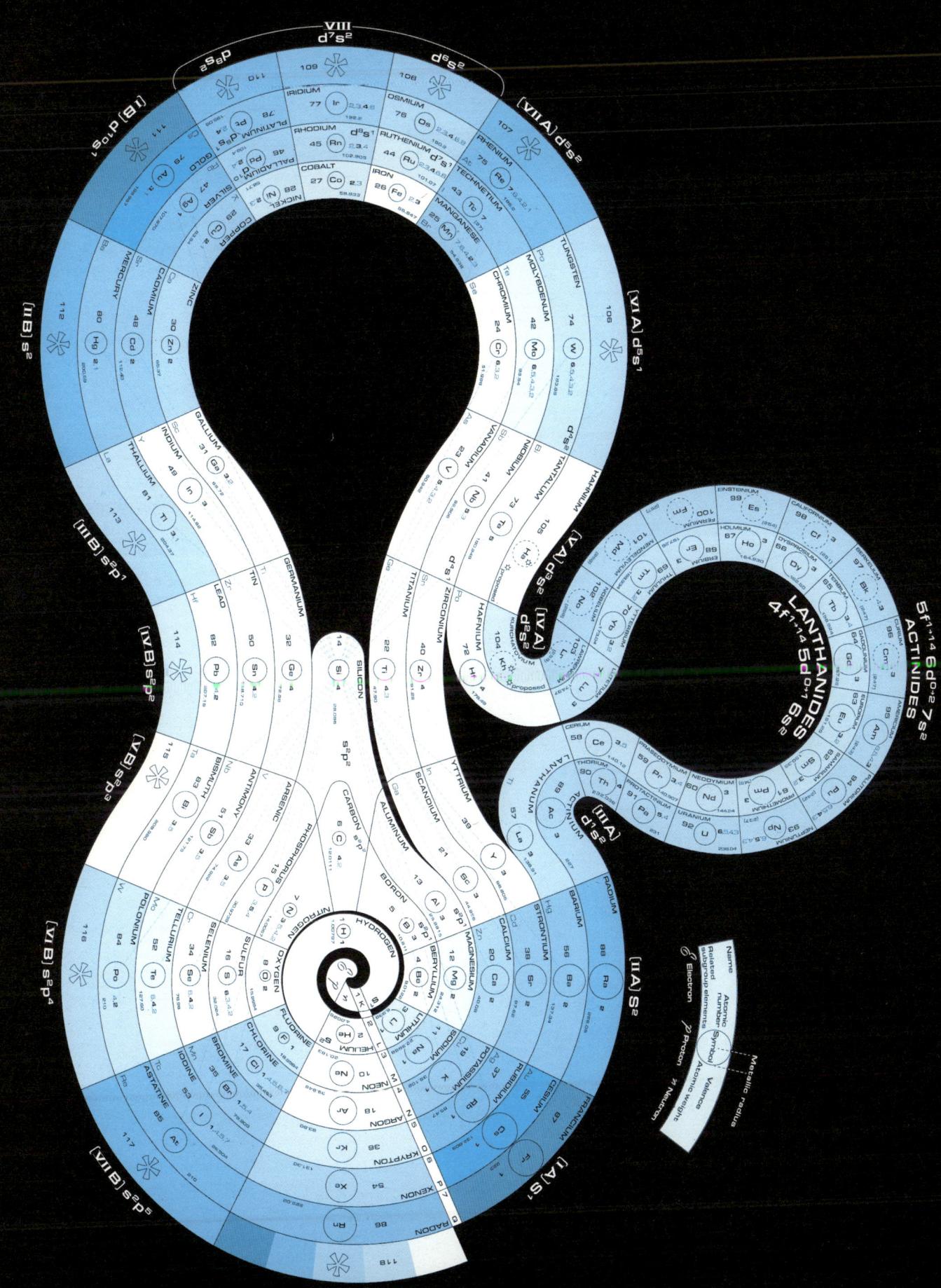

if he left some slots empty: elements presumably yet to be discovered. When some of these were found soon after and had just the properties he predicted, he was vindicated.

Still, it's a weird kind of periodicity. At first, chemical properties seemed to recur every eight elements. But in the row that starts with potassium, there's an interlude of 10 metals—the transition metals—and so it continues thereafter, creating a periodicity of 18. And after lanthanum (element 57), chemists discovered a whole series of 14 metallic elements with almost identical properties that have to be squeezed in too—frankly, these elements, called the lanthanides after the first of their ilk, all seem a bit redundant. There's another block like this after radioactive actinium (element 89), called the actinides. In most Periodic Tables, the lanthanide and actinide blocks are left floating freely underneath so the table doesn't get stretched beyond the confines of the page. (Some insist that this long-form table is the only proper one.) Why this odd structure?

The answer became clear with the invention of quantum mechanics in the early twentieth century. The chemical properties of elements are mostly determined by how the electrons in their atoms are arranged. New Zealander Ernest Rutherford showed that atoms comprise a central, very dense nucleus with a positive electrical charge, surrounded by enough negatively charged electrons to perfectly balance that charge. Rutherford imagined the electrons orbiting the nucleus like moons, but in the quantum-mechanical description they occupy nebulous, smeared-out clouds called orbitals. Using quantum mechanics to describe the disposition of electrons shows that they are arrayed in shells. The first of these can contain just two electrons—this is the only shell possessed by hydrogen and helium, the two lone elements at the tops of the towers—while the next has eight, and then 18. The shape of the periodic table thus encodes the character of the quantum atom.

All clear? Not quite. Even now, there's no consensus about how to draw the Periodic Table. Hydrogen, the first and lightest element, has always been awkward: it tends to get plonked on top of the first column (the alkali metals), but it doesn't really fit there—it's not a metal, after all. Some prefer to see it float freely above the rest, a hydrogen balloon over the edifice of elements. And representing the rather awkward nuances of the quantum shell structure in a two-dimensional diagram involves compromises, which have prompted the invention of all manner of ingenious alternatives to the traditional block format: spiral and circular tables, loops and stadium shapes, tiered ziggurats, three-dimensional models, dizzyingly imaginative cartographies of elements. None has caught on.

Some of the fiercest arguments involve the lanthanides and actinides, which begin in the third column of the Table. Which elements truly belong in those two slots? Older tables put lanthanum (symbol La) and actinium (Ac) there, with the rest of the series relegated to those disconnected basement blocks. Others instead assign those two positions to the last of the lanthanides and actinides: lutetium (Lu) and lawrencium (Lr). Some leave the position undefined, labeled only 'La–Lu' and 'Ac–Lr.' The problem is that the arguments for one choice are chemical—which elements are chemically most similar to scandium and yttrium higher up in column three?—while the other option is preferred quantum-mechanically, based on how the electrons are configured. In some ways this is a dispute about authority. Which has the final say on the Periodic Table: chemistry or physics?

Put that way, you can see the potential for acrimony. I was for a time a member of a group tasked by the International Union of Pure and Applied Chemistry—the authority on chemical nomenclature and systematization—to make recommendations that might resolve the matter. But the group couldn't agree, and so the argument continues. Or to give it a more positive spin: you're free to choose the Periodic Table you like best. ●

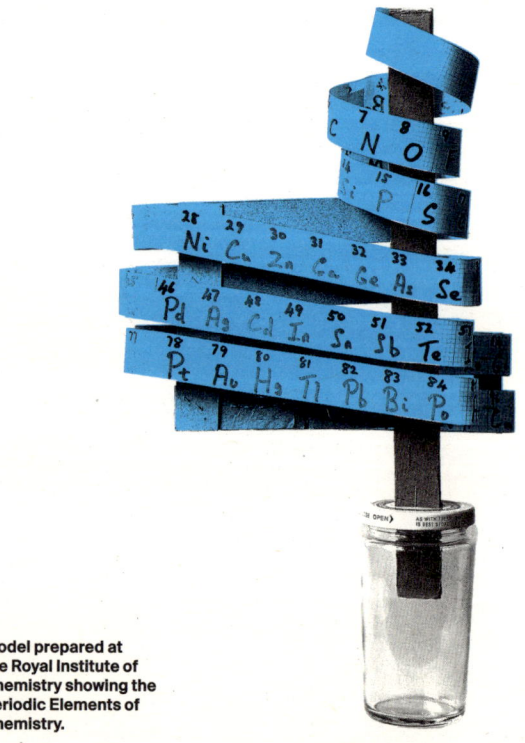

Model prepared at the Royal Institute of Chemistry showing the Periodic Elements of Chemistry.

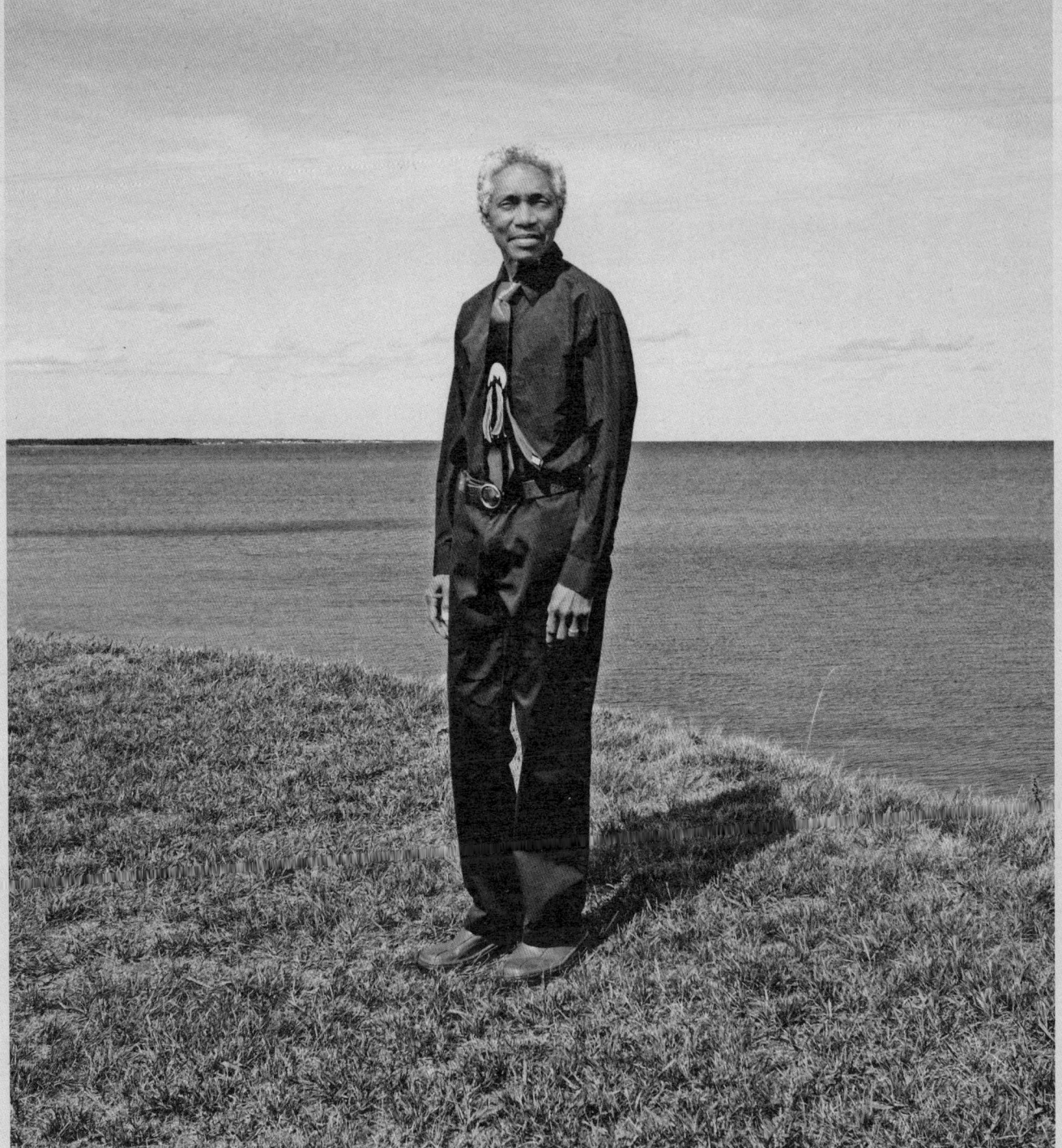

Beverly Glenn-Copeland

WE DIDN'T KNOW, WHEN THIS SHOW WAS BOOKED, THAT IT WOULD BE THE TREASURED ARTIST'S LAST IN NYC.

*K*eyboard Fantasies was passed down to me—a dub of a dub, as things sometimes were in the early 2000s—as an unmarked cassette in an unmarked case, slid into my hand after some house punk show. I'd never heard of Beverly Glenn-Copeland before, and what I adore about that album is that the experience of listening to it doesn't do much to reveal who Beverly Glenn-Copeland is. Glenn-Copeland is secondary to the atmosphere he's the architect of. The world you enter is his, but it could also be yours. It's lush, and ambient. When I first heard the tape, I told the pal who gave it to me that it sounded like outside but not even an outside I knew well. One I had maybe seen or felt in a dream. A green so vibrant, one can hardly look directly into it. This is the highest praise I can offer an artist: Glenn-Copeland is as generous a world-builder as there has ever been. One who, unlike our heavenly deities, builds and asks for nothing in return, except that you might be transported elsewhere for a little while. *—Hanif Abdurraqib*

I LIKE A PIECE OF FISH

JANNA LEVIN

My sisters and I were insulated from the Old World. We spoke un-accented English, unfettered by religious dogma. We were forward thinking, unburdened by the weight of tradition. But there was Grandma Eve, wrought out of nature's toughest minerals, a foreign substance from a brutal and primitive lost world. Eve was an enigma in some ways, an anathema in others, often unintentionally hilarious. For my newborn daughter and her toddler brother, I wanted a record of my impression of their complicated great-grandmother, knowing memory is fugitive, knowing that her incomprehensibility would deepen with the generations. I wrote this story a few years before Eve died in 2012. She was 101 when she passed—still lived alone, still wore heeled mules, still full of vinegar. A twenty-first century centenarian. Just a few months ago, my own mother passed, shocking I suppose because I wrongly presumed longevity would be her genetic inheritance.

Last Mother's Day was the first without my own, and while I mock faux holidays with indignation, May 12 induced its share of contemplation. As the date loomed, I was prompted to find this story and, with some hesitance, to share it here. All immigrant families are mystified by the other generations. Maybe you will find something relatable, even if not in the specifics. —*J.L.*

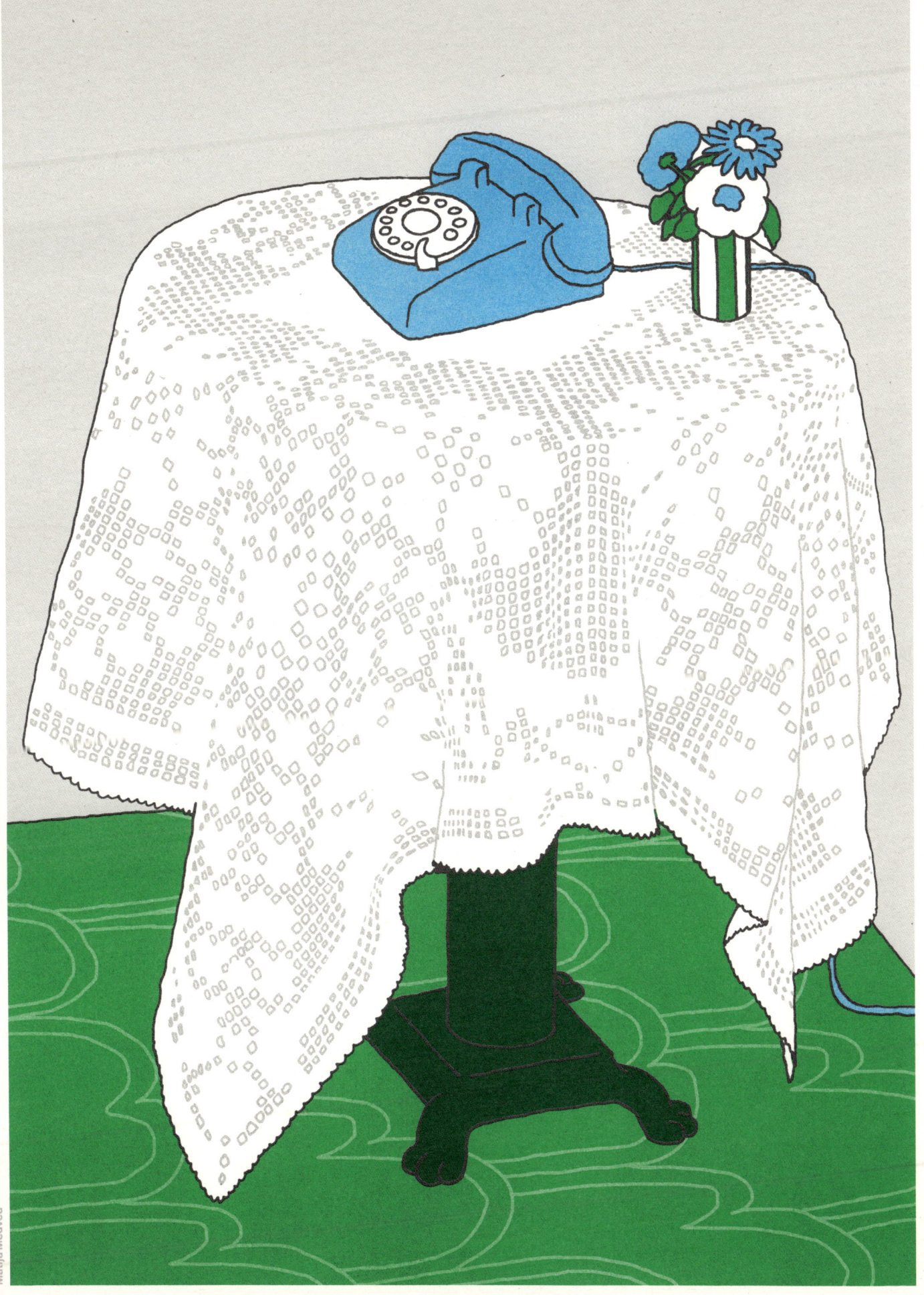

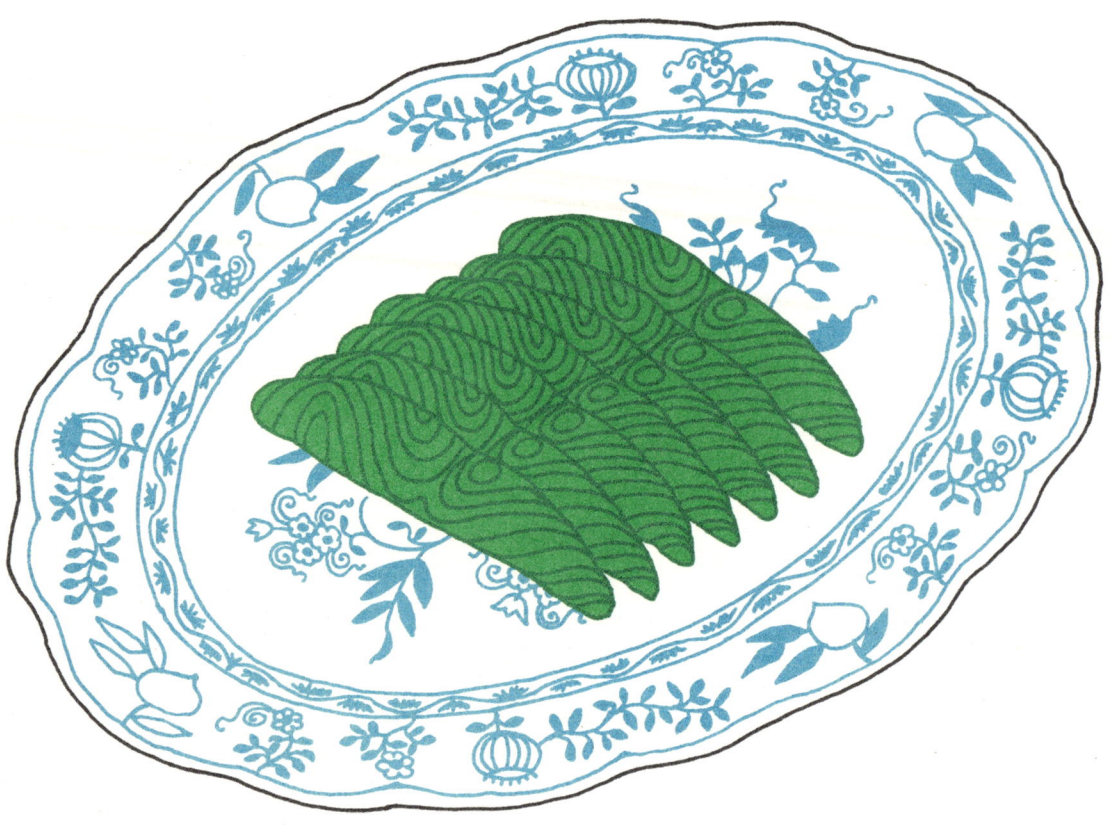

My 95 year-old grandmother can end almost any conversation with the words, "I like a piece of fish." When she phones, she ends with "a piece of fish," but she always begins with, "Did I wake you?" It could be 3:30 PM on a Monday. The phone rings and I hear, "Did I wake you?" She never stays on the phone long. I've started to take note of the time—six minutes usually—and I know she's about to hang up when she says, "I don't like to cook for myself. You can't make a big deal when it's just you." This last bit is a warning. "I'll defrost a little soup or defrost some chicken. I like a piece of fish."

When her great grandchildren were born, we switched from calling her "grandma" to calling her "Bubbie" or sometimes "Grandma Eve" for specificity. She has been deposed as the sole grandmother, but she is the only Bubbie. In Bubbie's apartment, defrosting is a magic process. Foods simply exist in her freezer. Before my sister Leslie and I drag our luggage through the door on our last visit together to her Chicago condo, she runs back into the kitchen to retrieve a frosty looking plastic sandwich bag.

"Look, I found kugel!"

She still runs around. She literally runs. The phone rings and she runs full throttle through the swinging saloon-style doors to the kitchen, she runs across the linoleum, she can't stop evenly so kind of bumpers into the little phone stand and knocks the dumbbell of a receiver off the landline's substantial black base. She usually does this running in open-back, high-heeled slippers. If my mother were there she'd shout, "Ma, don't run!" Bubbie always shouts back in Yiddish. Often while running.

Suddenly, Bubbie remembers the freezer, a self-standing, daunting and deep, rectangular icebox. She hurtles back toward the kitchen, lest the bounty should disappear. Standing on tiptoes to rifle through the icy contents, hinged at the waist, she disappears except for her legs.

"I found kanadlach! You want?… Poppyseed cookies!"

She runs back to us, eyes arched wide with wonder before tossing one of the zip-lock bags into the ancient microwave. She shrugs, "Ach, I must have made those when your cousins were here. I don't like to cook for myself. You can't make a big deal when it's just you. I'll defrost a little soup or defrost some chicken. I like a piece of fish."

When Bubbie was 86 she called to announce she had quit smoking. "Did I wake you? I quit smoking," she said with disappointment, her voice pure gravel, scarred from a lifetime of thin, brown, Mores cigarettes. "Ach, what are you gonna do?" Decades ago she had surgery on her vocal cords to remove the nodes, but her voice still resonates in a permanently smoky cavity. I have this image of her pumping her own gas with a brown cigarette hanging out of her mouth, clamped between teeth and moist with a sheen of pearly lipstick. It's an absolute fact that she always pumps her own gas, but the cigarette dangling above the fuel tank is probably an embellishment. "I have to go. I'm delivering Meals-On-Wheels to old people. We defrost some chicken, maybe a little soup. Me, I like a piece of fish." I have to ask, in disbelief, "How old are they'?"

I wonder if she drives the Meals-On-Wheels truck or just travels in the back with the tinfoil covered trays, talking to the other volunteers, gruffly laughing the way she does, making angry declarations about the world. But I don't have a chance to pose the improbable queries before she is off the phone. There are lots of questions I have for her. I need to ask, How did you escape Russia in a covered wagon? Why were you quarantined in Liverpool? How long were you on the boat before your brother died? I keep meaning to sit her down to get a careful account. I only know these few facts because I did ask her once, "Why did you leave Yumpala?" "It was the Russian Revolution!," she shouted.

My sister and I sit at the glass table in the kitchen with the ornate white-painted metal legs. I've known that table my whole life. We eat kugel and kanadlach and poppy seed cookies. Bubbie sits at an angle, toes straining to stretch her feet to the ground, resting her elbows on her knees and stabbing her powerful if tiny finger for emphasis while she lets us know her opinions on politics. "I'm a dumb old lady but I still know some things." Sometimes bits of masticated food fly out of her mouth while she talks.

Bubbie hasn't spoken to my mother for nearly two years. I try to discuss the rift with her on the phone sometimes but she gets so angry it sounds potentially fatal. "I'm an old lady! You can't upset me like this! It's not good for me!" I think she's started smoking again. It sounds like she's exhaling more than just breath into the receiver. I picture her literally smoldering with fury. When I get the chance, I'll check the drawers for cigarettes or lighters.

My mom tells me, "She never liked us. And we must have been so cute." They were cute. I've seen pictures of my mom and her identical twin sister when they were girls. Their hair descended in shiny spirals as they held hands in the faded photos of their matching outfits. They argue about who is who in the oldest pictures. They are mirror image twins down to their personalities. But it's hard to pick out these details in the pale images.

"I found lox!" My sister claims to have been in the other room when Bubbie found the lox in the freezer. We talk in the kitchen while she defrosts the lump of fish in the microwave. She stops a couple of times to massage the pink flesh between her hard fingers with thick nails arched high like calluses before throwing the aggregate—still wrapped in plastic and frozen in the center—back under high for 60 more seconds.

I watch Bubbie and my sister pick at the lox while I consume the poppyseed cookies. They are everyone's favorite cookie. Hard but not crumbly or dry. There must be 50 cookies in the clear plastic bag. I eat them cold. Not all of them, but a lot. They are never quite frozen, even when retrieved from the ice box.

I don't ever remember seeing her actually bake any cookies. They could always be found in her house but I never thought about how they got there. My father said he used to see her bake all the time when he was first dating mom and for decades after. He once saw her pulling tins of hot cookies out of the oven with bare hands. "Ma, let me help you," my dad approached, hands outstretched.

"DON'T TOUCH IT RICHARD!" Bubbie shrieked, "IT'S HOT!"

"How are you holding it?!" my father hollered, horrified, the burning metal tray in Eve's one hand as she warded him off with the other.

"I'M USED TO IT!"

After she laid the hot tray down on the counter, my father inspected her fingers for severe burns. They were slightly red but thick as leather gloves. Essentially undamaged. "Ma what are these marks on your forearms?"

"Ach, sometimes I burn myself," she shrugged. She rested the next set of hot trays on her arms as she slammed the oven door.

Her arms aren't burnt or red now. "I don't bake much since everyone's moved away," she explains. "Hmmm," my sister and I hum in sympathy.

Leslie plays with a small jewelry box on the kitchen counter like a teenager. "Take it," Bubbie insists. "It will remind you of me."

"Bubbie, why is this going to remind me of you? I've never seen it before."

"Take it!"

"Bubbie, I'm not going to take your jewelry box," my sister laughs and rolls her eyes.

"Take it!" Bubbie and I yell together. I know there's no point resisting. Sooner or later, Leslie's going to take the unfamiliar jewelry box to be reminded of her. Leslie throws her arm over Eve's shoulders. Eve pushes her tongue out slightly and winks.

"Here," she grabs my arm and we all go into her bedroom where she dumps jewelry out of another little box. "Janna, you take this. It will remind you of me."

"Bubbie, why is this going to remind me of you? I've never seen it before!"

"Take it!" Leslie shouts.

"How's your mother?" Bubbie musters finally, leaning an elbow on the counter by the spilled jewelry. "All the decent jewelry is gone already," she says. "What do I need to wear that stuff for? I'm not for a lot of stuff."

"Mom's good," we both say. Neither of us is inclined to embellish. She's great actually because she can't stand the lot of you and the silence is a profound release. That's how she put it: "The silence is a profound release!" We keep this part to ourselves.

"Good. That's all you can want for your kids. That they're healthy and happy. Even if they hate your guts, you wanna know they're good." She

yells this last part, arching her eyebrows as high as they would go. "Do you know what I mean, honey?" she says rhetorically, softening.

After we leave Bubbie in the doorway of her apartment on the 14th floor, Leslie and I swap jewelry boxes in the elevator. We look them over and they remind us of her.

The next morning, Leslie is crippled by food poisoning, I assume from the defrosted lox Bubbie found in the freezer behind the kugel and the poppy seed cookies.

"She defrosted the lox?" my sister implores, pleading with me to deny it when the first signs of nausea creep up. "Why did you let me eat it? Oh God!"

The image of the orange and yellow flesh twirling beneath the plastic wrapping between Bubbie's fingers is unforgettable. "Yeah, she defrosted the lox. I thought you knew. I was wondering why you were eating it."

"Oh God!" Leslie laughs loudly then moans, racing to the toilet.

While I lay on the huge, pillow-strewn hotel bed, barely able to hear my sister vomit and defecate in the luxurious hotel bathroom, I call Bubbie to make sure she is feeling okay.

"Bubbie, it's Janna. Did I wake you?"

"No, I wake up early. You don't sleep as well when you get old. I have oxygen sometimes. The tubes go all down the hall so I can still watch in the TV room. It's the local news. I watch all the current affairs shows. It's important! Do you watch news?" She has a dramatic, rising intonation. The television is incredibly loud in the background. "Honey, you need to stay informed! You better vote for Obama, not that McLain. Wait. Let me get my hearing aid."

"You mean McCain?"

"Ach, I can't hear a damn thing," she shouts when she comes back to the phone and then offers that gritty laugh. "Honey! When am I going to see you again?" We talk for six minutes, she sneaks in the part

about liking a piece of fish, and then we hang up.

Determining that she is fine and not poisoned, I tell Leslie through the bathroom door. "Bubbie's fine. She's indestructible." Finding this unbearably funny, Leslie howls painfully, a laugh-cry, the next wave of nausea part of the punch line.

When she can manage, she shouts through the door, "Bubbie ate most of the lox! I only had a small piece." Leslie's cackle is unhesitating. Poison lox is funny. She laughs deeply until she throws up again.

She stays in the bathroom for about an hour. Occasionally she whines apologetically, "I'm sorry."

"It's okay, Les," I whine back.

It's nearly two years before I visit Bubbie again. She answers the door in her gold-braided, high-heeled sandals. She hasn't changed noticeably since I last saw her in the doorway on the 14th floor of Winston Towers. It's really the 13th floor but the elevator buttons claim that the 12th floor is succeeded immediately by the 14th.

We talk around the glass table with the ornate white-painted metal legs. She tells me, "I'm the only one of my contemporaries left. The others in the building are much younger. They're like 80. I don't tell them how old I am. Half of them can't remember anyway. And I don't want to make them feel bad. They all use walkers except for me. The doctor, he sees me, he blew over. He shouts, '*You're 97! Look at your hair!*' My hair, shit! My hair looks terrible. But he expects a walker, a nurse—not hair. He nearly blew over." Her hair, remarkably, is salt and pepper, not white.

"Are you going to a lot of funerals?"

"What am I going to a funeral for? The coffee? Jesus Christ. Shit."

"Yeah. The coffee. Maybe the food too."

"I like some ribs, a bit of steak."

I wait for it. I even try to lead her. "But what do you really like?"

"Ach, I'll eat anything. Shit." ●

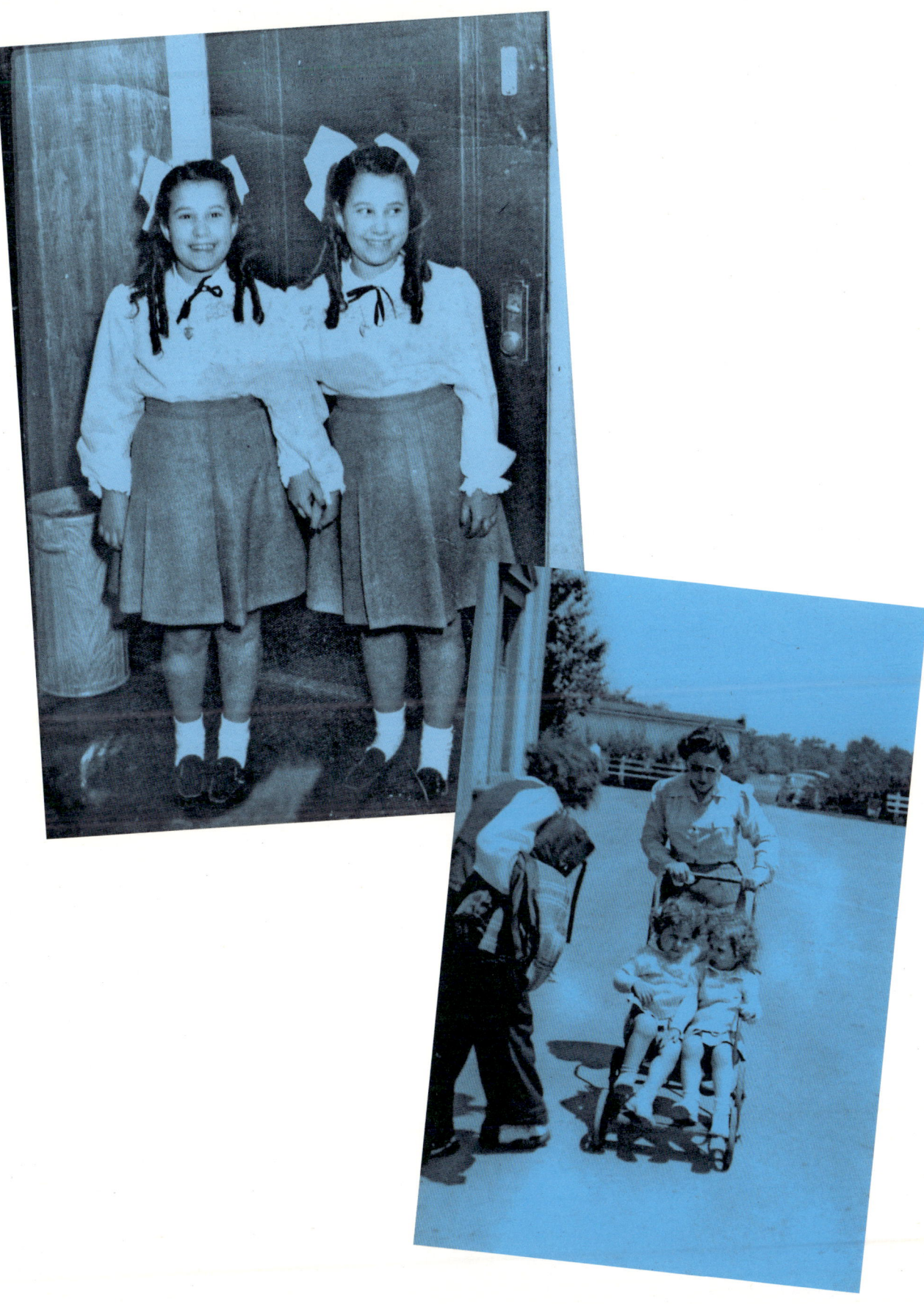

Above: Janna Levin's mother and aunt. Photo courtesy of Levin.
Below: Bubble and the twins. Courtesy of Denise Kavin

CONTRIBUTORS

HANIF ABDURRAQIB is a poet, essayist, and cultural critic from the east side of Columbus, Ohio. He is the author, most recently, of *A Little Devil in America* and *There's Always This Year: On Basketball and Ascension*.

HILTON ALS is a longtime staff writer at *The New Yorker*, where his criticism won a Pulitzer Prize in 2017, and an advisor to *Pioneer Works Broadcast*. He is a professor at UC-Berkeley and Columbia, and his books include *The Women*, *White Girls*, and *My Pinup*.

RAE ARMANTROUT is one of the founding members of the West Coast group of Language Poets. She is the author of more than ten collections of poetry, including *Finalists*, *Conjure*, and *Versed*, which won the 2010 Pulitzer Prize in Poetry, a 2009 National Book Critics Circle Award, and was a finalist for the 2009 National Book Award.

PHILIP BALL is a science writer who trained as a chemist at the University of Oxford and as a physicist at the University of Bristol. His work appears in *Nature* and *Chemistry World*, and his books include *Curiosity*, *Critical Mass*, *The Modern Myths*, and most recently, *How Life Works*.

BRAD BOLMAN is a Postdoctoral Member in the School of Historical Studies at the Institute for Advanced Study. He is the author of *Lab Dog: What Global Science Owes American Beagles* (forthcoming from the University of Chicago Press) and is currently at work on a new book, *Rotten Beauty*, about the often deadly roles of fungi in human history.

NATASHA BOYD is a writer from Los Angeles. She has contributed to *The Nation*, *The Drift*, the *LA Review of Books*, *Artillery,* and elsewhere.

SWAMP DOGG is an American singer, musician, songwriter, and record producer. He debuted his new soul sound on the *Total Destruction to Your Mind* album in 1970, and released an LP of country songs with Joyful Noise and Pioneer Works Press, *Sorry You Couldn't Make It*, in 2020. He is the subject of a new documentary, *Swamp Dogg Gets His Pool Painted*, and the author of *If You Can Kill It I Can Cook It*.

MEGAN FERNANDES is a South Asian American writer living in New York City. She is the author of *The Kingdom and After* and *Good Boys*. Her third book of poetry, *I Do Everything I'm Told*, was published in 2023.

VIVIEN GOLDMAN is a writer, broadcaster, educator, and musician. Her most recent two books are *Revenge of the She-Punks: A Feminist Music History from Poly Styrene to Pussy Riot* and *The Book of Exodus: The Making and Meaning of Bob Marley and the Wailers' Album of the Century*. An anthology of her journalism, *Rebel Musix, Scribe On A Vibe*, is out this year.

JOSHUA JELLY-SCHAPIRO is Director of Publishing at Pioneer Works and co-editor-in-chief of *Pioneer Works Broadcast*. His books include *Names of New York* and *Island People*, and he teaches at NYU.

DANIEL KOLITZ is a Brooklyn-based writer. He has contributed to *The New York Times Magazine*, *The New Republic*, *The Atlantic*, and *The Nation*, among other publications.

JANNA LEVIN is the Founding Director of Sciences at Pioneer Works and the co-editor-in-chief of *Pioneer Works Broadcast*. She is a professor of physics and astronomy at Barnard College of Columbia University. Her most recent book is *Black Hole Survival Guide*.

ISLE MCELROY is a Brooklyn-based writer and the author of two novels: *The Atmospherians* and *People Collide*. Their other writing appears in *The New York Times*, *New York Times Magazine*, *The Guardian*, *The Cut*, *Vulture*, *GQ*, *Vogue*, *The Atlantic*, and elsewhere.

ELVIA WILK is a writer and editor living in New York. She is the author of the novel *Oval* and the essay collection *Death by Landscape*.

REBECCA WRAGG SYKES is an archaeologist, author, and public scholar. She is an Honorary Research Associate and Honorary Fellow at the Universities of Cambridge and Liverpool. Her widely acclaimed first book *Kindred: Neanderthal Life, Love, Death and Art* won the 2021 PEN Hessell-Tiltman prize for history.

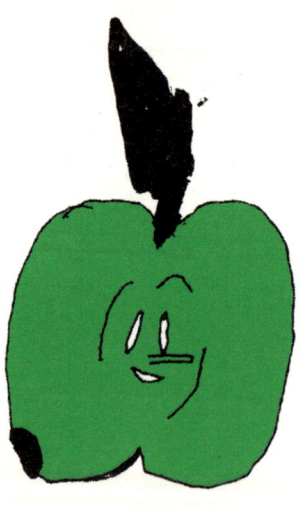

ILLUSTRATIONS

Callum Abbott (pp. 31-36, 59, 61, 66, 73, 78, 81, 83); Eric Reh (pp. 65, 89, 93, 105); Luc Tedford (pp. 63, 92); Trevor Davis (pp. 72, 74, 77); Matija Medved (pp. 100-101); Chi Park (p. 88)

PHOTO CREDITS

Mycellenz/Wikimedia Commons/ CC-BY-SA-4.0 (p. 10), (© Michael Wood, courtesy of MykoWeb) (pp.11-17); Joshua Jelly-Schapiro (p. 20-21); Landesmuseum Württemberg, Stuttgart, Germany/ Wikimedia Commons/CC0 1.0 (p. 25, 29-30); Wellcome Images, London, United Kingdom/ Wikimedia Commons/CC0 1.0 (p. 27-28); Kent County Council, Walter (Jo) Ahmet, Kent, United Kingdom/Wikimedia Commons/ CC0 1.0 (p. 23-30) Sugarloaf scrapbook, courtesy of Joan Heckenberg (pp. 37-39, 44, 43-45, 48); All photos courtesy of Swamp Dogg, with the following exceptions: (pp.49-57); Lionel Hampton (William P. Gottlieb/Library of Congress/PD) (p.54); George Jones (Pictorial Press Ltd/Alamy (p.55); Reproduced by Jeremy Sachs with permission from George and Sylvia Schuster (p. 96); Wellcome Images/Wikimedia Commons/CC-BY-4.0 (p. 97); Beverly Glenn-Copeland. Photo: Paul Atwood (p. 98); Janna Levin and Denise Kavin (p. 104)

TYPEFACES

JHA Times Now, Times Ten, and Zipper. Featuring: TF Burko Gorpo, ITC Benguiat Gothic, Roslindale, ITC Fat Face, Compacta, Cooper, Cheee, Microgramma, Churchward Tua, Manuka Slab, Roberta, Strike, Windsor Antique, Belwe, Heroe Pro, Croissant, Taters, Stentor, Wonder, Churchward Marianna, TwoBlock, TF Margate, MD Polychrome, MD Nichrome, Choc, Roslindale, Quaint, Tiny, WT Zaft, Media FS, Megazoid, Mercurius, and Kenwyn.

COVER ART

Promotional image for Alexandra Bachzetsis's *Escape Act* (2018), performed at Pioneer Works on April 11-12, 2019. Pictured: Tamar Kisch. Photo: Blommers & Schumm